Things Maps Don't Tell Us

Things Maps Don't Tell Us

AN ADVENTURE INTO
MAP INTERPRETATION

Armin K. Lobeck

With a new Foreword by Mark Monmonier

THE UNIVERSITY OF CHICAGO PRESS
Chicago and London

This edition is reprinted by arrangement with Macmillan
Publishing Company, a division of Macmillan, Inc.

The University of Chicago Press, Chicago 60637
The University of Chicago Press, Ltd., London

ISBN 0-226-48877-2 (pbk.)

Library of Congress Cataloging-in-Publication Data

Lobeck, A. K. (Armin Kohl), 1886–1958.
    Things maps don't tell us : an adventure into map
interpretation / Armin K. Lobeck.
        p.   cm.
    Originally published: New York: Macmillan, 1956.
    Includes index.
    1. Physical geography.   2. Geophysics.   3. Map
reading.   I. Title.
GB59.L6   1993   92-44623
910'.02—dc20   CIP

⊛ The paper used in this publication meets the minimum
requirements of the American National Standard for
Information Sciences—Permanence of Paper for Printed
Library Materials, ANSI Z39.48-1984.

TABLE OF CONTENTS

IV. LAKES

FOREWORD TO THE 1993 EDITION

I am gratified that the University of Chicago Press is reprinting Armin K. Lobeck's *Things Maps Don't Tell Us*. The book is a classic and deserves to be made available again to geographers, geologists, students of natural history, and thoughtful travelers. Lobeck's words and pictures hold unique insights and understanding for anyone curious about plains, coasts, mountains, and rivers. He not only helped me understand many places I had visited and taken for granted but also encouraged me to visit and appreciate many others.

Lobeck was a gifted cartographic illustrator. He understood the land surface as well as anyone of his era, he knew how to draw and interpret maps, and he mastered the concise integration of maps, block diagrams, and written text. The author of a once important college textbook on geomorphology, Lobeck made his broad knowledge and appreciation of the physical landscape accessible to a larger audience through *Things Maps Don't Tell Us*.

Born in New York City, in 1886, Lobeck had an influential career as an illustrator, geology teacher, and geographer. He studied architecture and botany before earning a Ph.D. in geology. In addition to a distinguished teaching career at the University of Wisconsin and Columbia University, he served the State and War Departments as a geographer and illustrator during World Wars I and II. His book *Block Diagrams*, published in 1924, exhibits many of the techniques that give his work a unique style. A successful entrepreneur as well as a talented author and artist, Lobeck founded the Geographical Press and published several informatively detailed physiographic maps and many other instructional aids.

Although advances in "process geomorphology" have rendered Lobeck's textbook on geomorphology hopelessly obsolete, *Things Maps Don't Tell Us* offers a unique and insightful introduction to the origins of landforms. Interpretations based largely on geologic structure and macro-level geomorphic processes explain its continued vitality. Although advances in plate tectonics and theories of continental drift now account quite nicely for faults and other structural features, Lobeck takes these structures as given and presents a valid interpretation of landforms that developed in response to continental plates, sea-floor spreading, and other geological features and processes. Similarly, his interpretations transcend more contemporary experimental and theoretical explanations for wind erosion, long-shore drift, glaciation, and other geomorphic processes.

Most readers can overlook the book's few conceptual flaws, which largely reflect the obsolete theory of "erosion cycles," whereby landscapes evolve through discernible stages of youth, maturity, and old age, usually to be reborn again through renewed tectonic upheaval. Lobeck's chronological explanation of Appalachian elevations, drainage patterns, and water gaps (pp. 108–113) and his discussion of the "peneplains" of southern Sweden (p. 149) draw heavily on the now discredited paradigms of William Morris Davis and other early-twentieth-century geomorphologists. In the early 1960s, geologists realized that geologic structure, weathering, hill-slope movement, and fluvial erosion and deposition provided a "dynamic equilibrium" that made the erosion cycle and its hypothetical peneplains wholly unnecessary. Although my own 1967 master's thesis was one more nail in the Davisian coffin, I nonetheless admire Lobeck for his concise explanation and demonstration of the peneplain concept.

Things Maps Don't Tell Us is a collection of cartographic detective stories. Its mysteries are the coastlines, rivers, lakes, and other commonplace physical features that map users normally take for granted. After pointing out patterns that call for solutions, Lobeck assembles clues and leads his reader through thoughtful, carefully crafted explanations. Gratified and enlightened by these discoveries, the reader is left with a richer appreciation of what maps can tell us if we care to look and listen.

Mark Monmonier
Syracuse, New York

INTRODUCTION

Maps tell us a great deal. But there are some things which maps don't tell us, interesting things too. By this I mean that there are some facts hidden on the map for us to read if we know how to do so.

Most atlases and collections of maps which we have in our homes show simply the shapes of continents and countries, their shore lines and islands, rivers and lakes, political divisions and places, and occasionally cities and towns with roads and railroads. Some atlas maps have also green and brown tints to show elevations above the sea.

Most of us, I dare say, are so accustomed to taking all these things for granted that we never in our minds raise any question as to why the coast lines, islands, rivers, lakes, and other geographical features have the shapes and the locations which they possess. The explanations for some of these things can easily be guessed. The explanations for others are less obvious. Therefore I am going to give you an opportunity to discover some of these things. On the LEFT-HAND pages of this book I am setting before you a *problem* in the form of a simple map. This map shows a few geographical details which I have taken from one of the maps, atlases, or road maps that you yourselves are likely to have at hand. I have also called attention in the accompanying text to a few of the details of the map which it would be well to puzzle over a bit before referring to the page on the RIGHT to see what the *explanation* actually is. Thus, I hope, your curiosity will be piqued and you may be brought to wonder why things have come to be as they are. You will be right, too, if you come to the conclusion that there is a meaning for almost every fact represented on the map, tens of thousands of them, and not only the few score which are presented in this book.

As we study the maps of the different continents, our attention is directed to several categories of features. Foremost, of course, are the COAST LINES, the details of which determine the shapes of the continents. We note on the one hand the various promontories, peninsulas, capes, points, and headlands that project outward from the coasts. And, on the other hand, we observe the many and varied gulfs, bays, seas, estuaries or straits that indent the coast, in some cases permitting the ocean to extend far inland.

And next, closely allied to the coastal features, are the ISLANDS of the world. Some are contiguous with the mainlands, and others lie in the remoter parts of the seas. Their variety in shape, size, and location poses many questions.

Then there are the RIVERS of the world. Rarely do we raise any question about their relation to one another and to other features, nor do we give much thought to their many curves and irregularities. Nevertheless, the story behind these details is often an absorbing one.

Almost a part of the river systems are the LAKES of the world in their many varieties and shapes. The explanations for lakes alone would encompass almost the entire range of geological phenomena, the same phenomena which under slightly different circumstances explain the shapes of shore lines and rivers. Like a theme with variations, we find these different geological events occurring again and again under slightly different settings, thus accounting for the apparent infinite variety of detail that appears on our maps.

Finally, we may wish to mention a few man-made details such as roads, railroads, and CITY PLANS. Occasionally we shall find that even in these cases some past geological event has been influential.

Map Reading and Map Interpretation. The subtitle to this book is "AN ADVENTURE INTO MAP IN-TERPRETATION." Let me therefore say a few words here about what map interpretation is. In order to do so, I must first tell you what map reading is, and then explain the difference between map reading and map interpretation.

Map reading is what all of us do when we want to find out where a place is on the map, or the distance between places, or their relative positions, or any other simple geographical fact. Map reading is what boy scouts do when they use maps out of doors to keep from getting lost. Map reading is what the motorist does when he uses a road map to find the best route for his journey.

But map interpretation is much more than all this. Map interpretation is like the process of reading between the lines of a story whereby the reader draws certain inferences and conclusions which the author did not specifically make. On a map, for example, the form of the hills, the location and pattern of rivers, the outline of the coast, and other details may provide the clue to certain geological information which the map makes no pretense of giving. So, in this book, we shall take certain simple geographical information which we can *read* from the map, and see what inferences we can draw from it, to *interpret* and explain how these details have come about.

I have said that this book is intended to be an Adventure in map interpretation. The subject of map interpretation is a fairly scientific one, and is used mainly in the study of the detailed large-scaled topographic sheets of the various countries of the world. A vast amount of information can be inferred from these maps, much more, indeed, than the maps actually indicate. Here, in this book, however, we have much more limited material with which to work. It is necessary therefore to confine our efforts, and not attempt to cover the entire subject. We shall do merely what we can by consulting maps that are readily available, maps which of one kind or another we are apt to have in our own small personal libraries. These would include such school atlases as *Goode's School Atlas*, the new *Bartholomew Advanced Atlas of Modern Geography*, Rand McNally's and Hammond's atlases, and also similar maps to be found in the encyclopedias. It would include the automobile road maps of the various states and the splendid maps issued by the National Geographic Society.

Although this book is completely self-sufficient, and does not require reference to any other maps, it is presumed that those really interested in maps will look up the actual areas described herein on one of the maps mentioned above.

Things Maps Don't Tell Us

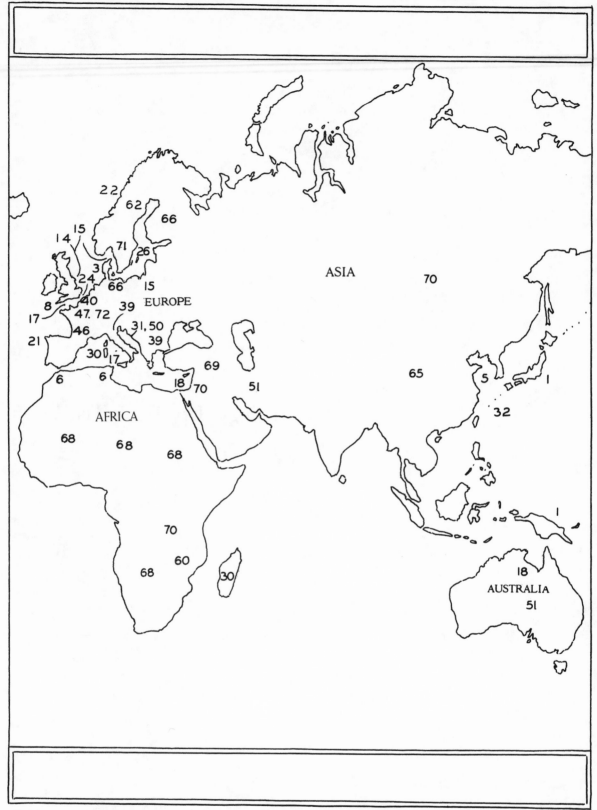

The numbers on this map indicate the location of the "Examples" mapped or described on the following pages.

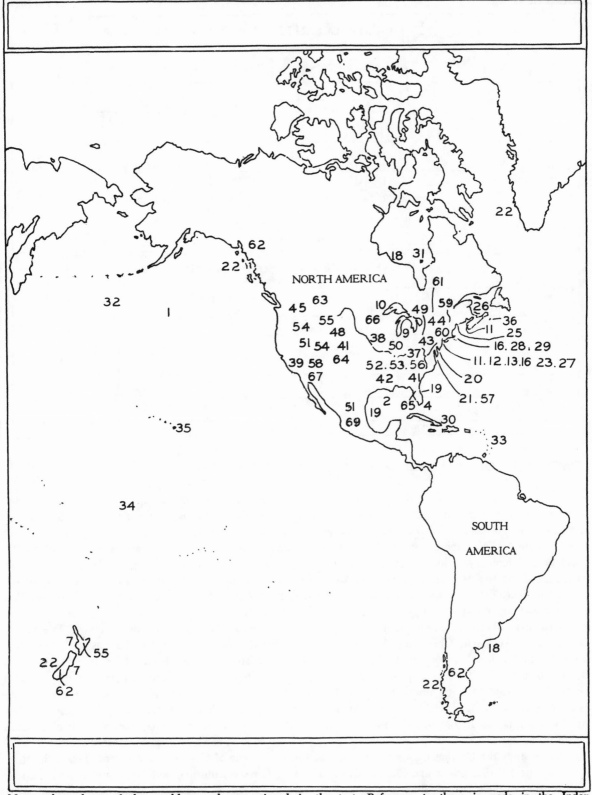

Many other places of the world are also mentioned in the text. Reference to these is made in the Index.

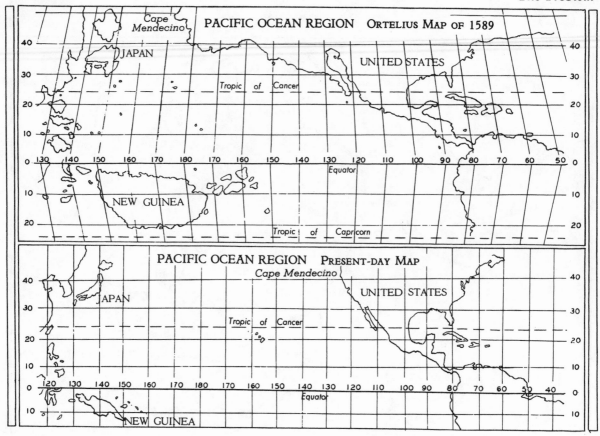

COAST LINES. *Distorted Coast Lines. A Map of the World of 1589.*

Most of us are so accustomed to seeing continents depicted in a certain way upon the map that any distortions of their outlines promptly attract our attention. We look at old maps with considerable amusement and with a certain feeling of superiority. We could probably draw the map better than that from memory, we are apt to think. To examine a series of old maps is to see mankind groping for an understanding of the world around him. The older maps start off falteringly, the earliest ones omitting whole continents and ignoring entirely the Western Hemisphere. Gradually, as explorations carried men to the far corners of the world, knowledge increased and more details were added; nevertheless, for many centuries the relative location of places was poorly understood. The resulting maps were therefore distorted.

Take as a rather typical example the map drawn by Ortelius in 1589 and reproduced above. The Pacific Ocean was known. Japan, China, and the East Indies were familiar places. South America was settled and had reached a high cultural development, the University of San Marcos in Lima having been founded in 1551. But the maps of that time were terribly out of shape. The western coast of North America, as may be seen above, was thought to reach almost to Japan. The island of New Guinea looks like a continent, and reaches as far east as California. Japan is in the right place but badly out of shape. Yucatan extends out eastward, instead of pointing north. Beneath the 1589 map shown above is a correct present-day map of the Pacific Ocean region for comparison.

As we study the Ortelius map we note the equator, the Tropic of Cancer, and other parallels of latitude. And we note too that places like Florida, Cuba, Yucatan, Lower California, and Japan are located properly with relation to these latitude lines.

The longitude lines have been renumbered on the above Ortelius map to conform with the prime meridian of Greenwich, as the original map was based upon Spain. Longitudinally, points on the Ortelius map are very much out of position. This is the problem which demands an explanation.

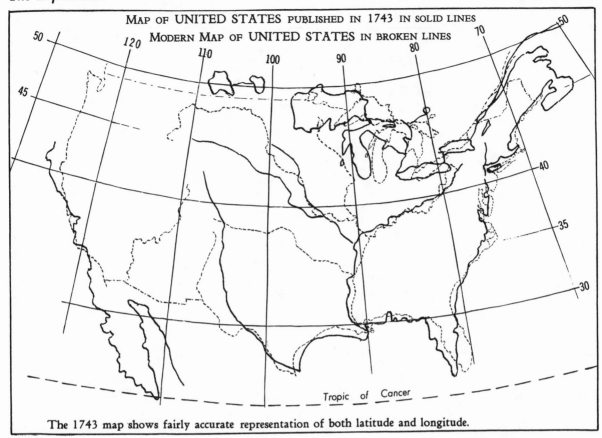

MAP OF UNITED STATES PUBLISHED IN 1743 IN SOLID LINES
MODERN MAP OF UNITED STATES IN BROKEN LINES

Tropic of Cancer

The 1743 map shows fairly accurate representation of both latitude and longitude.

The improvement of maps through the centuries resulted primarily from two causes. First, the greater knowledge of the world added to the store of accurate information that could be put on a map, and made it unnecessary for map makers to invent geographical features with which to fill in the blank spaces. In the second place, it became possible to determine the longitude of places with greater and greater precision.

The determination of latitude was always fairly easy and accurate, especially in the Northern Hemisphere. The latitude of a place is always the same as the altitude of the North Star above the horizon, as seen from that place. In the Southern Hemisphere it was not quite so simple; but a knowledge of the movement of the sun and its elevation above the horizon at different seasons at different places made the determination of latitude fairly accurate almost anywhere on the surface of the globe.

Longitude, being measured from some prime meridian, is always the number of degrees of the earth's circumference east or west from that meridian. The size of the earth being known, it is easy to determine the number of miles in a degree of longitude at different latitudes on the earth's surface. On land these miles can be measured off with some accuracy, but at sea winds and currents render this almost impossible. Inasmuch as it apparently takes the sun 24 hours to go around the earth, this means that the sun's apparent path covers 15 degrees every hour. Therefore if we know the difference in time between two places it is easy to determine their difference in longitude. Nowadays this is easily done by means of telegraph and radio, or even by means of accurate chronometers. Before the advent of good chronometers, however, during the 18th century, longitude was determined by crude methods. Even a fairly good clock, losing not more than 10 seconds a day, would after a 2 months' voyage be 10 minutes out; this would represent an error of 2½ degrees, or some 170 miles, in the equatorial regions. It was no uncommon thing for clocks to be 2 hours off after a long ocean trip; this resulted in an error in longitude determination of 30 degrees, or over 2,000 miles, thus easily explaining the distortions on old maps. The above map of 1743 shows the improvement that came about, during the 18th century, over that of the Ortelius map of 1589.

3

EXAMPLE 2   *The Problem*

COAST LINES. *Projections and Protuberances. Gulf of Mexico.*

To begin our examination of coast lines, we shall discuss first the shores of the Gulf of Mexico. Here we note several projections and protuberances. Some are large and irregular, as along the coast of Louisiana. Others are mere outbowings of the coast, hardly detectable on the map, such as we find along the coast of Texas. Other coast lines of the world have similar outbowings, as for example the African coast of Nigeria, and of Egypt, and the east coast of India. It is worth while to look at any atlas map of these regions and note their general similarity. There are many other examples, too, if we care to search for them.

Almost more noticeable than the outbendings of the coast are the indentations. Indentations are to be noted along the coast of Texas. They occur along all of the other Gulf states as well.

Then there are also the long offshore islands, the so-called barrier islands, particularly noticeable off Texas. Features like these will receive special consideration later in this book.

Most noteworthy of all the features shown on the above map is the large peninsula of Florida. It would look strange indeed if Florida were omitted from the map. Nevertheless it would be more normal, or let us say more ordinary, if Florida were not there. Florida is actually an exceptional feature, worthy of a special explanation. This explanation has been reserved for Example No. 4.

The Gulf of Mexico is very much like the Mediterranean Sea and other large enclosed bodies of water. The small features along such shores closely resemble each other. That is to say, these shores all have small outbowings or irregular projections. On the other hand, the larger features in the two regions are totally different. Florida, for example, does not resemble Italy at all. Southern Europe, in other words, is very different indeed in geological character from the southern United States. This means, therefore, that the larger features of the coasts are different. The smaller details of the coasts, however, in the two regions, are much more alike. The larger features are part of the continental framework, whereas the smaller details are influenced by the nature of the enclosed seas, which in the two regions is similar.

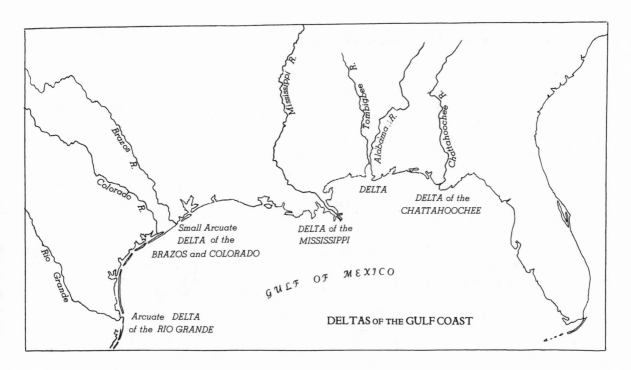

DELTAS OF THE GULF COAST

On the above map the chief rivers of the Gulf Coast region have been indicated. This enables us to see clearly that most of the protuberances and irregular projections are closely related to the rivers, and are in fact caused by the rivers. They are all river deltas. Let us examine several of them.

The Mississippi delta is the most conspicuous. This great delta has been formed by one of the mightiest rivers of the world. The Mississippi River carries to the Gulf of Mexico each year more than 400,000,000 tons of sediment, or more than 1,000,000 tons a day. This represents a mass sufficient to cover 1 square mile to a depth of 268 feet. Much of this vast amount of silt drifts far out into the Gulf. But tremendous quantities settle down close to the mouth of the river. In enclosed seas like the Gulf of Mexico, alongshore currents are weaker than in the open ocean, with the result that sediment is not carried very far from the mouth of the stream.

The Mississippi River is constantly extending its mouth into the Gulf of Mexico. As a matter of fact the Mississippi River has several mouths. That is to say, it has several distributaries and its delta therefore has several branches. This kind of delta is known as a bird's-foot delta. It is one of the less common types. Another small example of such a delta is that built by the St. Clair River, the outlet of Lake Huron, into Lake St. Clair, near Detroit.

The commonest type of delta is the arcuate delta. This is simply an outbowing of the coast. On the map above, it is clear that the Rio Grande by its deposits has gradually pushed the coast line outward into the Gulf. The Brazos and the Colorado of Texas, too, have definitely modified the form of the shore line. So also have the Alabama-Tombigbee, as well as the Chattahoochee.

In relatively recent times a slight rising of the waters of the Gulf of Mexico has caused a submergence of the mouths of the smaller streams, and numerous small embayments have resulted. The larger streams have more than kept pace with this submergence, and by delta-building have avoided the formation of estuaries at their mouths.

5

COAST LINES. *Peninsulas. The Peninsula of Denmark.*

The peninsula of Denmark projects north from the continent of Europe toward the much larger peninsula of Scandinavia. It forms an effective barrier between the Baltic and the North seas. Indeed, it holds such a commanding position that the Danes for centuries were able to exact tolls from vessels passing through the narrow channels of the Skagerrak and the Kattegat.

There are some noteworthy facts about the Danish Peninsula which may possibly throw some light upon its origin and help to explain some of its unique characteristics. The accompanying map, below, shows (as any atlas map would) that Denmark is not only a long peninsula but that it is made up also, on its eastern side, of many estuaries and irregular islands of all shapes and sizes. Note also that the eastern shore line of the peninsula is very irregular, whereas the western side of the peninsula, toward the North Sea, has a more continuous and uninterrupted outline. If you have a fairly large-scale atlas map before you, one which shows by colored tints the height of land, you will notice too that the eastern parts of the peninsula, as well as the larger islands, are several hundred feet higher than the low western plain bordering the North Sea. You will

note also the presence of occasional small lakes and ponds among the eastern hills similar to the much larger and more numerous ones which dot the entire North German Plain adjacent to the Baltic Sea.

As our attention is turned toward North Germany, we note also that its northern coast, like Denmark, is somewhat irregular, with such islands as Rügen. And we should note also the island of Bornholm, which is actually an outlying member of the Danish archipelago.

This little section of Scandinavia, with its helter-skelter of features, is not to be taken for granted. It has had an interesting history, and has come about because of certain well understood natural events.

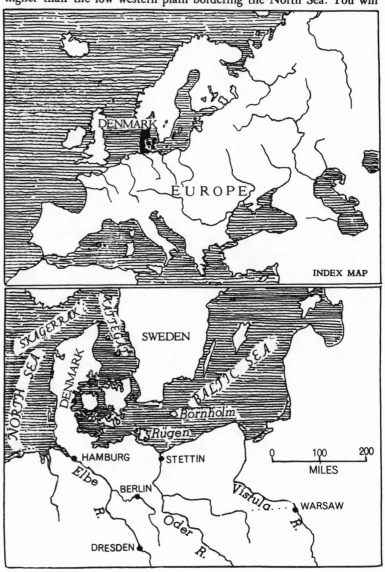

The peninsula of Denmark, as shown by the maps below, is part of the terminal moraine of the continental ice sheet. Figure 1 shows a great lobe of the continental glacier pushing its way south from the Scandinavian Highlands which lie to the north. Standing in this position for many years, and constantly feeding forward its load of debris, the glacier built up its terminal moraine in the form of numerous hills and hollows. Just outside the moraine, to the south, and to the west, an outwash plain was formed, deposited there by the melting waters from the ice sheet.

Rivers that, prior to the advance of the ice, flowed north to the Baltic, like the Vistula and the Oder, were blocked by the advancing ice and forced to turn west into the Elbe and thence to the North Sea. In Figure 2 we see the same region after the disappearance of the ice. The morainal hills are now represented by the many islands and irregularities of eastern Denmark and northern Germany. In Denmark the moraine stands in the water, whereas in North Germany the moraine stands on dry land. The sandy outwash plain forms the low sand plains of the North German Lowland, as well as the flat land of western Denmark. The old channels of the Vistula and the Oder remain, and serve as routes for some of the great canal systems of northern. Germany. Berlin and other cities are located along these old channel ways, which all lead west to the Elbe River and eventually to Hamburg. For this reason Hamburg is able to serve as the port for much of eastern Germany and of Poland. These inland canals therefore obviate the need of shipping through Baltic ports, which would entail a much longer journey by sea.

Were we to look elsewhere for a counterpart of Denmark, we would find it on Long Island. There the higher northern part consists of morainal hills, whereas the southern part is a low sloping outwash plain.

THE PENINSULA OF DENMARK

EXAMPLE 4   *The Problem*

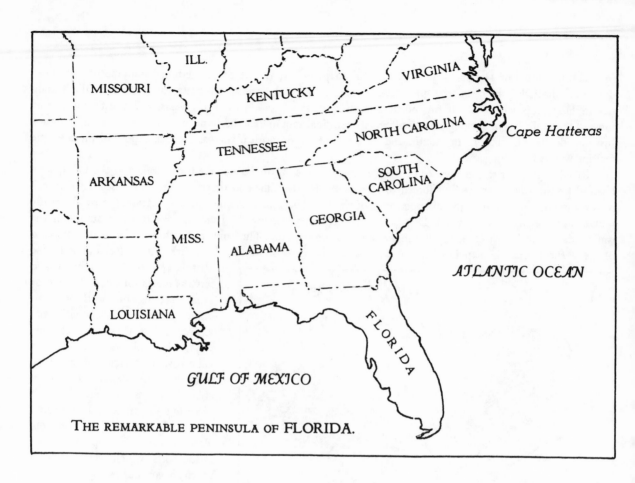

THE REMARKABLE PENINSULA OF FLORIDA.

COAST LINES. *Peninsulas. The Peninsula of Florida.*

Like a great thumb, the peninsula of Florida projects from the southeastern United States for a distance of 400 miles. No other feature of the nation's coast line is so prominent. Without Florida a map of the United States would look queer indeed. So prominent is Florida that every schoolchild knows its location almost before he learns the locality of any other state. People across the seas, to whom many of our states are unknown by name, are sure to know the name and the whereabouts of Florida. And in the history of our country Florida has had a unique importance.

Compare Florida with some of the other great peninsulas of the world. It is about the same size as Korea, but not so irregular in shape. On the continent of Europe, Italy most nearly resembles Florida. It has about the same width but is twice as long. In addition it is much less regular because of its "toe" and its "heel," as well as its "spur." Neither Africa nor South America has any peninsulas remotely resembling Florida, nor does Australia. Florida is indeed a unique and unusual feature with no counterpart anywhere else in the world. Its nearest geological relative is actually right on the Atlantic seaboard itself, just north of Florida. Notice the bulge in the coast line where North Carolina projects into the ocean at Cape Hatteras. This bulge is simply the beginning of another peninsula like Florida, strange as that may seem.

Unlike Example No. 2 in this book, Florida is not a delta. Nor is it anything at all like Denmark. Perhaps it is more like Yucatan, its close neighbor. Both of these peninsulas, Florida and Yucatan, actually have much in common.

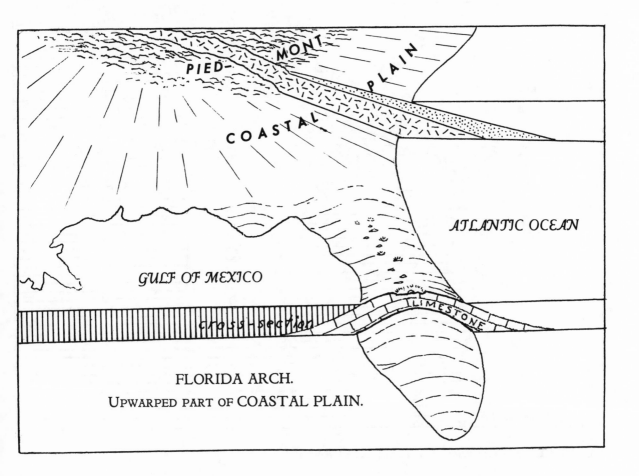

Florida is an uparched portion of the Atlantic Coastal Plain. It is a low anticline. The rest of the Coastal Plain slopes off seaward into the shallow water overlying the continental shelf. If any part of the continental shelf were uplifted or arched up, as Florida is, it would become an island. If near the mainland, it would become a peninsula like Florida, long or short, depending upon the length of the arch. If very short, it would be just a large bulge like the outward sweep of the shore at Cape Hatteras. In western Florida there is another slight bulge which projects into the Gulf at Apalachicola. The above diagrammatic map illustrates most of these facts, and shows by cross section the structure of Florida and the rest of the Coastal Plain.

In Florida most of the beds of rock which have been arched up are limestone. One series of limestone beds is known from borings to be 500 feet thick. Where exposed at the surface, and where not covered over by sand and clay deposits, the limestone is eroded into caves and sinkholes. There is much underground drainage, and the streams frequently come to the surface in great springs. Many of the depressions thus formed hold lakes, thousands of them of all sizes. They make up a belt which runs north and south through the center of the state, along the axis of the uplift. In some places the limestone is rich with the bones of marine animals, and is an important source of phosphate. Toward the south the limestone beds slope beneath the level grasslands of the Everglades, and disappear almost imperceptibly into the Gulf.

Except for Yucatan, which resembles Florida in several ways, there is no other large peninsula that has been formed in like manner.

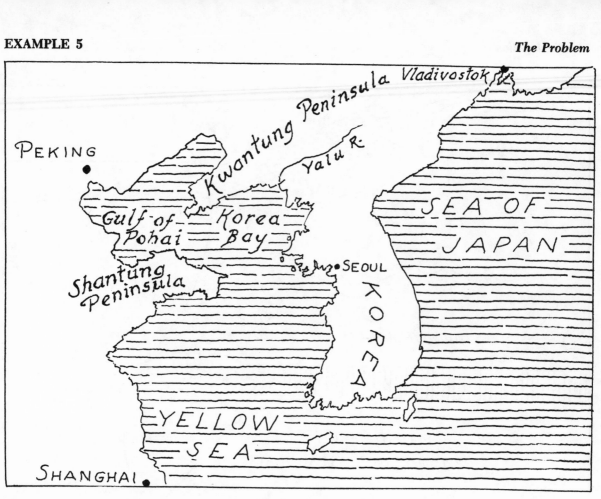

COAST LINES. *Peninsulas. The Peninsula of Korea.*

Only slightly larger than Florida in size is the peninsula of Korea. At first glance it is not so much different in shape, either. And yet it has had quite a different origin. As we study Korea on the map, let us examine the Kwantung Peninsula and the Shantung Peninsula at the same time. These three peninsulas are all related to one another.

A perusal of the above map brings out the fact that the peninsulas and gulfs which surround the Yellow Sea and the Sea of Japan have a fairly well marked angularity. And we note also that several of the straight-line features trend in northeast-southwest direction. Not only here but in much of eastern Asia do we find that this is true.

Inasmuch as the straight-line features of eastern Asia were formed many ages ago in geological time, some of their original sharp-sided characteristics have been lost as a result of stream and wave erosion. They are actually mere wrecks of their former clean-cut selves. Some outstanding features, however, still remain. You will note, for example, that the straight shore line forming the southern side of Kwantung Peninsula, along Korea Bay, continues on toward the northeast without change of direction along the course of the Yalu River. And this same straight line may be carried on toward the southwest to mark out some of the features in China, as we shall see. As we examine the map still further, we notice several other straight-line features which seem to be continuations of other similar features elsewhere on the map. Not only do the larger features fall into this pattern, but innumerable small details of the coast line also have this angularity. The river systems of eastern Asia, especially the smaller tributaries, likewise show a tendency toward an angular arrangement, so much so indeed as to produce a checkerboard-like layout of the topography.

These straight-line phenomena just enumerated serve as the clue to the explanation for many of the geographical features of eastern Asia.

10

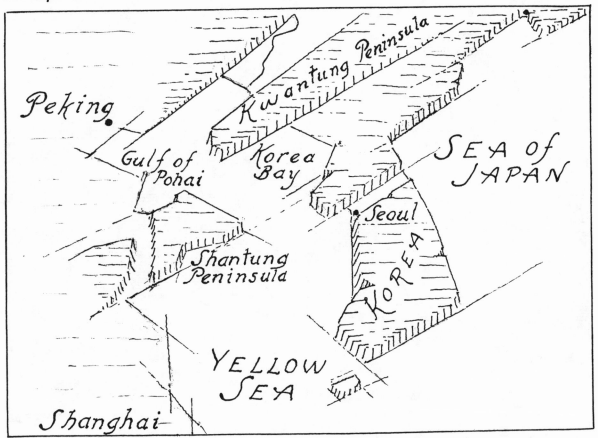

The above diagrammatic map emphasizes the straight-sided outline of Korea as well as the regions adjacent to the Yellow Sea and the Sea of Japan. This illustration may seem like a fanciful exaggeration and an effort to read into the map more than is actually there. As a matter of fact the picture is not overdone very much. It is in fact somewhat simplified. There is always some difficulty in detecting angular patterns of geographical features because erosion by rivers and the cutting action of waves along the coast disfigure the original sharp outlines. Simplicity gives way to complexity of form. Many little irregularities and details are introduced, so that in time the angular pattern is obscured or even obliterated.

Now, as to how these forms came about. This part of the world, like many others, as we shall see later, was subjected to pressure. The crust of the earth was not wrinkled and folded, as it was in many places, to form mountain ranges, but instead was cracked and broken. The cracking and breaking of a large mass of brittle rock like this usually takes place along straight lines, the lines being often at right angles to each other. The tendency to form a rectangular pattern can readily be demonstrated by laboratory experiments on glass and such brittle materials.

The blocks produced by this breaking were then disjointed and uplifted or tilted in different directions. A displacement that occurs along a crack of this sort is called a "fault," whereas if the rock is only cracked or "fractured" it is called a "joint."

A field examination of the whole Korean region illustrated above would reveal millions of fairly closely spaced joints or cracks in the rock outcrops. If measured and mapped, it would be seen that these small fractures would fall in line with the larger elements of the landscape. It is in fact these small observable phenomena which help us to understand how the larger features have come about.

EXAMPLE 6   *The Problem*

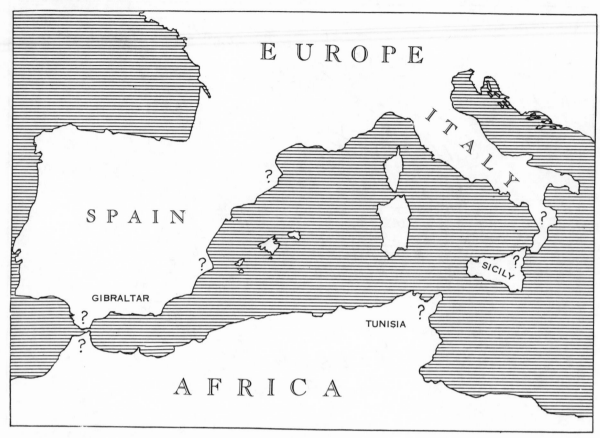

COAST LINES. *Peninsulas. The Peninsula of Tunisia, Africa.*

The peninsula of Tunisia in North Africa, which is shown on the map above, is quite different in shape from the peninsulas of Korea and Florida described in the last two examples. The most striking element about Tunisia is its irregular outline, and particularly the long projection which points toward the island of Sicily.

Around the shores of the western Mediterranean are several other knobby projections which also seem to require an explanation. From Spain on the north and from Africa on the south are two projections which are directed toward each other and which are separated by the narrow Strait of Gibraltar. Along the Mediterranean coast of Spain are several other protuberances, the problem of whose origin may likewise be of interest. To these several promontories a question mark has been placed on the map. Italy and Sicily too have several noteworthy projections or peninsulas. Particularly notable are the "toe" and the "heel" of Italy. The toe of Italy points toward the long eastern point of Sicily, from which it is separated by the very narrow Strait of Messina. The whirlpools and strong currents in this strait were feared by the ancient sailors, and gave rise to the legend of Scylla and Charybdis, two female monsters with heads of fierce dogs who were a menace to seafarers.

It would seem that if we could explain any one of these points and protuberances, we might perhaps at the same time find a clue to the others.

In the western Mediterranean two groups of islands may also be pointed out. The western group of three or four small islands constitutes the Balearic Isles. The two large islands of the eastern group are the islands of Corsica and Sardinia. The contrasting form of the islands in the two groups is worthy of attention. The Balearic Isles with their small projecting points are quite different from the bold angular outline of Corsica and Sardinia. In Example 30 the origin of these large islands will be explained, while perhaps the Balearics will fit into the scheme to be presented on the following page, a scheme which will also explain Tunisia.

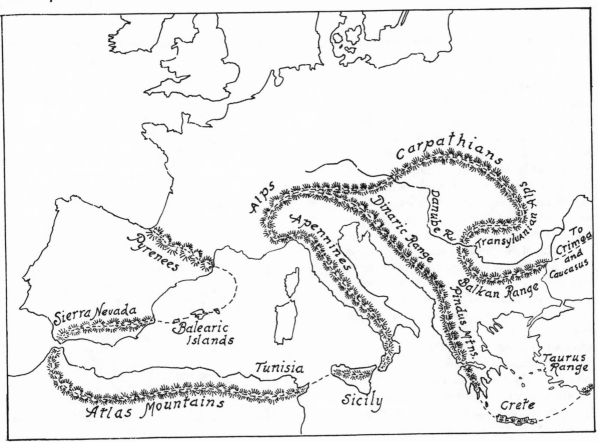

The dominating features, as well as many of the smaller details of southern Europe and of North Africa, are determined by the so-called Alpine System of mountain ranges. The several members of this mountain system are indicated on the above map. All of these ranges have been produced by the folding and crumpling of the earth's crust. The result is that each unit or range consists of several parallel chains or ridges, much more complicated than has been suggested on the relatively simple map shown above.

Turning our attention now to some of the details, we may start first with the Alps themselves, from which the whole Alpine System gets its name. The Alps form the border lands between Switzerland, France, and Italy. They include some of the highest peaks in Europe, arranged along several more or less well defined chains. One of these chains swings around toward the south to form the Apennines of Italy. In southern Italy this chain turns toward the west along the toe of Italy, then through Sicily, and enters Africa through the peninsula of Tunisia. This peninsula constitutes the eastern end of the Atlas Mountains. This range in the west turns north and crosses into Spain at the Strait of Gibraltar. Continuing in southern Spain as the Sierra Nevada, the range then passes into the Mediterranean where its broken crest forms the Balearic Isles. Beyond these islands there is a wide break before the range reappears in the Pyrenees between Spain and France.

The eastern part of the Alpine System hardly concerns us here, but it will be interesting to mention its different members. Eastward from the Alps we note that the Alpine System separates into two parts. One part swings around to the north to form the Carpathians, the Transylvanian Alps, the Balkan Range, and then crosses the Black Sea to reappear in Crimea and the Caucasus. The southern branch forms the Dinaric Range of Yugoslavia, and the Pindus Mountains of Greece. Beyond Greece, this chain passes into the island of Crete, and then forms the Taurus Range in Asia Minor.

EXAMPLE 7   *The Problem*

COAST LINES. *Peninsulas. Taranaki Peninsula, North Island, New Zealand. Banks Peninsula, South Island, New Zealand.*

New Zealand consists of two main islands, known as North Island and South Island. North Island is by far the more irregular of the two. It has several projections or peninsulas, each having its own peculiar shape. Taranaki Peninsula is quite definitely the simplest of them all. It sticks out on the western side of North Island like a huge bump. The explanation for Taranaki is evidently quite different from that of the other peninsulas. The two peninsulas that make up the eastern half of North Island are actually parts of a long more or less continuous range of mountains which goes by the name of the Tararua Range in the south and the Raukumara Range in the north. The long peninsula north of Aukland, with its many estuaries, is in a special category of its own.

South Island is long and relatively narrow, and not too complicated in outline. It trends in a northeast-southwest direction and thus falls right into line with the two ranges which make

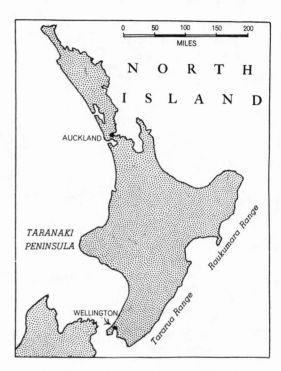

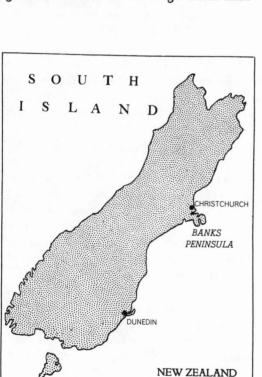

up the southern part of North Island. It may be fairly safe to assume that a long mountain range forms the main axis and backbone of South Island. There is, however, one little protuberance which seems out of place. It is Banks Peninsula, near Christchurch. Along this otherwise simple coast, this button-like projection surely seems anomalous. Note that its outline is not a complete circle, but is indented by two embayments which run almost to its center. On a coast otherwise poorly provided with harbors, these embayments are important. On the northern one, which extends inland ten miles or more, is situated the busy little city of Lyttelton (not shown on the map) which is connected by rail with, and serves as the port for, Christchurch. On the southern embayment is the port of Akaroa (not shown either), which is connected by highway with the splendid road system of the Canterbury Plains nearby.

Taranaki Peninsula on North Island, New Zealand, is formed by one of the larger volcanoes of the world. Its name is Mount Egmont. Standing in the path of the prevailing westerlies, its 8,260-foot summit is almost constantly shrouded in clouds. On the plains around its base there are daily intermittent showers with the result that this part of New Zealand is endowed with splendid pastures. It supports a prosperous dairy industry which is one of the mainstays of New Zealand economy.

Mount Egmont, however, is only one of a number of volcanic peaks which dominate the landscape of North Island. In the central part of North Island there is a volcanic area which in several respects resembles Yellowstone Park. It is essentially a plateau made up of volcanic deposits, mostly volcanic ash which produces a white and dusty landscape. Two great peaks rise above this plateau, much as Mount Washburn does in Yellowstone Park, overlooking Yellow-

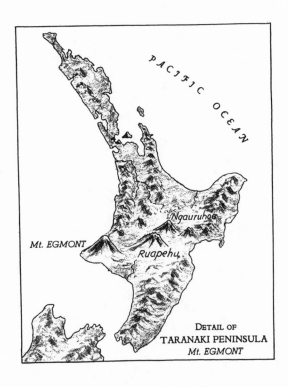

DETAIL OF
TARANAKI PENINSULA
Mt. EGMONT

stone Canyon. Ruapahu, the taller of the two peaks, is over 9,000 feet in elevation. Ngauruhoe, the more frequently active peak, is over 7,000 feet. There are many smaller peaks and cones, as well as a geyser and hot-springs region in this volcanic area of North Island.

Banks Peninsula on South Island is also a volcano, but quite different from those on North Island. As the adjacent sketch shows, Banks Peninsula is now only the stump or wreck of a former high peak which blew its top off. This left a large depression, or caldera, much like a gigantic crater. It is exactly like the caldera in which Crater Lake, Oregon, lies, but much larger. The embayments mentioned on the preceding page are places where the sea has been able to reach into the heart of the caldera in much the same manner as it has done in some of the islands of the Azores. The resulting harbors are protected on almost all sides. The port of Lyttelton is connected with Christchurch by a railroad which tunnels through the caldera rim.

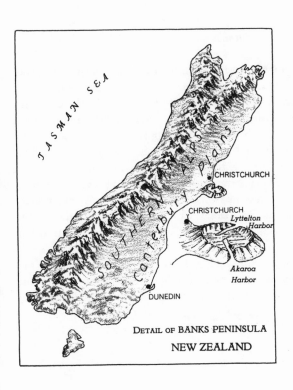

DETAIL OF BANKS PENINSULA
NEW ZEALAND

15

EXAMPLE 8

The Problem

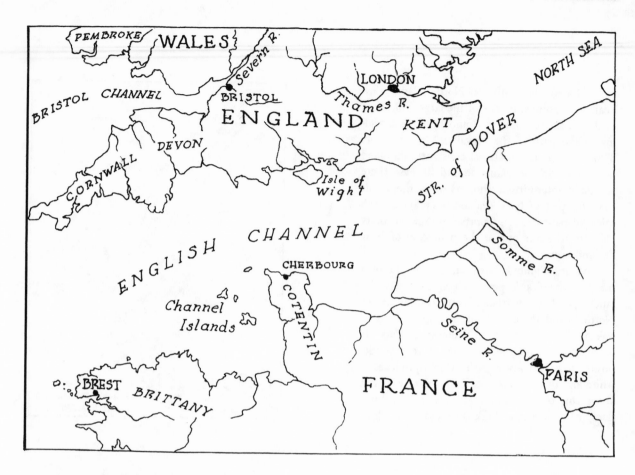

COAST LINES. *Peninsulas. The Peninsula of Cornwall, England.*

The peninsula of Cornwall, which includes not only Cornwall but the County of Devon as well, is but one of the many irregularities along the coast line of the British Isles, or for that matter of the entire continent of Europe. In southern England, farther to the east, is the peninsula of Kent, southeast of London. In Wales, north of Cornwall, is the Pembroke Peninsula; and south in France, across the Channel, is the Brittany Peninsula, and also the smaller Cotentin Peninsula, where Cherbourg is situated. We observe not only these large peninsulas but we note also many smaller peninsulas projecting from the margins of the larger ones. And we note also certain islands, quite close to shore, which are almost peninsulas. Such is the Isle of Wight, and such also is the island of Anglesea in northern Wales. The Channel Islands and many smaller ones too could also be mentioned.

Between the various peninsulas are many embayments: large embayments between the large peninsulas, and small embayments between the small peninsulas. For example, between Wales and Cornwall lies the Bristol Channel. Between Cornwall and Brittany lies the English Channel, which narrows at its eastern end into the Straits of Dover. Between the Brittany Peninsula and the Cotentin Peninsula is a broad and shallow embayment in the center of which are the Channel Islands. Into these embayments, both large and small, flow many rivers: the Severn, the Thames, the Seine, the Somme, and numerous smaller ones less well known. This pattern of promontories, embayments, and rivers is carried around the whole coast line of Europe and in greater or less degree around all the continents of the world.

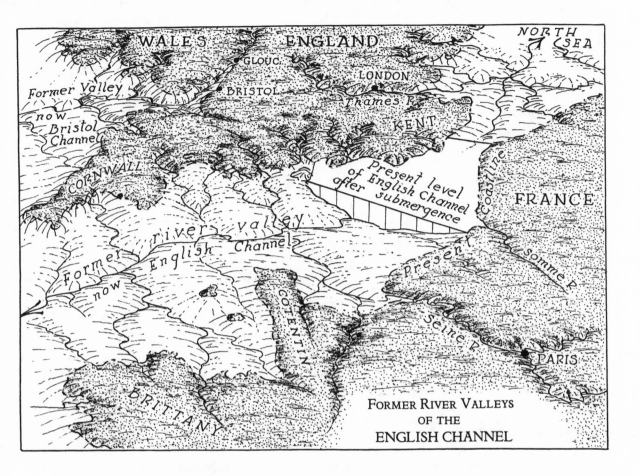

FORMER RIVER VALLEYS
OF THE
ENGLISH CHANNEL

Many of the peninsulas and embayments which occur so abundantly throughout the world are due to the sinking of the land or to the rising of the sea. Or, to put it another way, they are the result of a partial submergence of the coast line.

The embayments that have just been mentioned on the previous page are submerged or "drowned" river valleys. The sketch above is intended to suggest the appearance of the English Channel region prior to the encroachment of the sea upon the land. Where the English Channel now lies there was formerly the valley of a large river, with its many tributaries. The embayment now occupied by the Bristol Channel was formerly the extension of the Severn River valley. East of the Straits of Dover was another large river which flowed toward the northeast into the North Sea. Tributary to it was the Thames, and also the Rhine farther east. All of these river valleys have become embayed or submerged. The present position of the coast line, and therefore the shape of England as it appears on the map, is almost fortuitous, depending as it does largely upon the amount of submergence. If there had been much more submergence, the peninsula of Cornwall might have disappeared entirely. The English Channel and the Bristol Channel would together have covered it. The Isle of Wight has become detached from the mainland because the drowning of the area has completely submerged the streams which surrounded it.

The mouths of the Thames and of the Severn are drowned rivers, as are also the mouths of most of the rivers of England.

EXAMPLE 9

The Problem

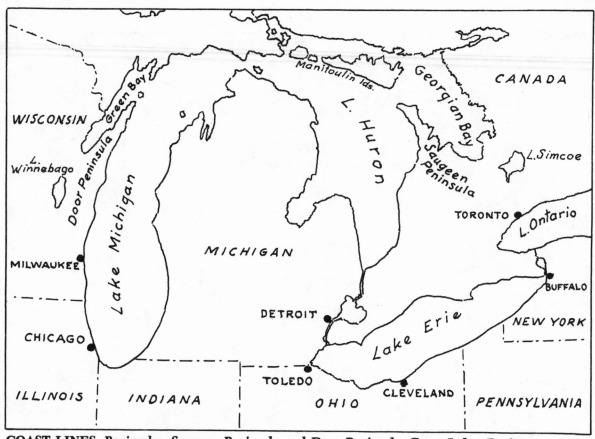

COAST LINES. *Peninsulas. Saugeen Peninsula and Door Peninsula. Great Lakes Region.*

And now we have two relatively small peninsulas which are identically alike in origin and quite similar in shape and size. They are the Saugeen, or Indian, Peninsula, which lies between Lake Huron and Georgian Bay; and the Door Peninsula of Wisconsin, which lies between Lake Michigan and Green Bay. Topographic features of this sort occur throughout the world, but only occasionally do we find them forming peninsulas. Sometimes they form islands, but usually they occur inland from the coast.

Notice how these two peninsulas, together with various islands and mainland features, form almost a perfect circle around Lakes Huron and Michigan. In fact, as we look more closely at the map, we notice that the state of Michigan is almost completely encircled by these two Great Lakes. It is interesting to note how the various features in the eastern half of the area are balanced by similar features in the western half of the area. Let us list them as follows: (*a*) Lake Michigan is balanced by Lake Huron. (*b*) Door Peninsula corresponds with Saugeen (Indian) Peninsula. (*c*) Green Bay corresponds with Georgian Bay. (*d*) Lake Winnebago in Wisconsin corresponds with Lake Simcoe. (*e*) There are even certain little features (islands and a peninsula) in the west which correspond with the Manitoulin Islands in the east.

These comparisons are mentioned here because, as we shall see on the next page, they help to bring out the similar origin of the two members of each pair. All of the various features just mentioned are so intimately related that if we explain any one of them we explain them all. For instance, an understanding of how Saugeen Peninsula came about will also explain why Lake Michigan is there, as well as all of the other details. We shall find throughout this book that an explanation of some small detail will very likely provide an understanding for many of the other details in the same region as well. Not only that, but once we have the clue to an understanding of certain things in one part of the world we very likely will find that this is also the clue to an understanding to certain things in another part of the world also. As a matter of fact, however, the Great Lakes region is almost unique. Only one other region of the world can be closely compared with it; namely, the Baltic region of northern Europe. But there is no room here now for this interesting story.

18

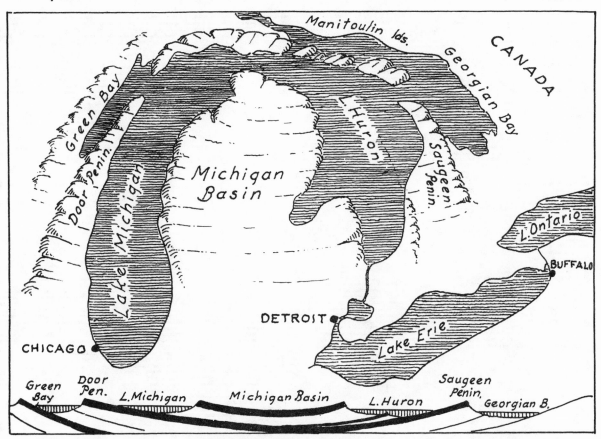

The circular pattern of the features in this region indicates that its geological structure is that of a "basin." This particular structure is known as the "Michigan Basin." The region is made up of sedimentary layers that are arranged like a series of plates or saucers piled one on top of another, with the largest at the bottom. The cross section below the sketch map above shows what we mean. The lakes and other water bodies occupy the lower belts between the edges of the saucers. For example, on the above map, we see that the edge of the saucer which produces the Saugeen Peninsula also produces the Door Peninsula, and similar bodies of water lie in the depressions on either side of these two peninsulas.

The Michigan Basin actually is only one of a series of domes and basins which make up the Great Lakes region. Erosion of these several structures has left the ridge-like features, or cuestas, standing higher than the intervening lower belts where the beds are weaker and therefore more readily eroded. After these circular lowlands were formed by various stream systems, the whole area was overrun by the continental ice sheet. When the ice melted, a great deal of debris, masses of irregular soil or glacial till, was deposited. This till blocked up many of the valleys, so that these lowlands became partially filled with water and lakes were formed. A similar history explains certain similar features in the Baltic region of Europe, such as the Gulf of Finland, Lakes Onega and Ladoga, and the White Sea of the Arctic.

Did you ever wonder, when you looked at a map of the Great Lakes, about the shape of Lake Huron? The northern part of this lake is much wider across than is the southern end. The diagram above shows why this is so. The northern half of Lake Huron covers two lowland belts as well as the intervening small (unnamed) cuesta. The southern end of the lake occupies only one lowland.

Doubtless as you study the above diagram you will try to explain a number of other things. How about Lakes Erie and Ontario? Why are they there? The answer is that they fit right in with the continuation of this story to the east, but which it is possible here only to hint at.

19

EXAMPLE 10   *The Problem*

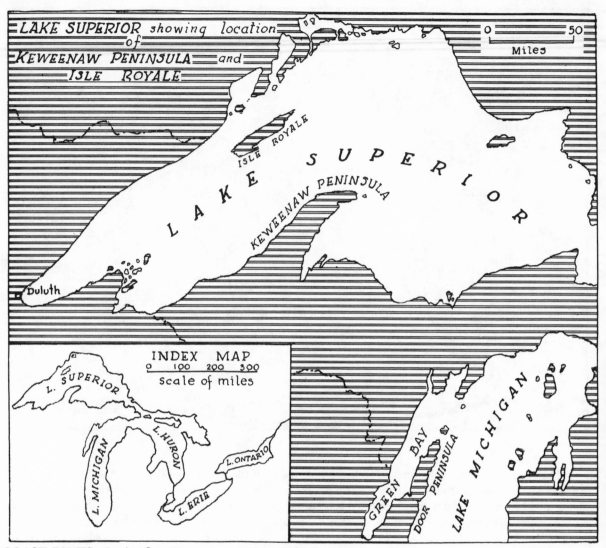

COAST LINES. *Peninsulas. Keweenaw Point, Lake Superior.*

On the two preceding pages it was shown that the Door Peninsula of Wisconsin and the Saugeen Peninsula of southern Michigan, together with the Manitoulin Islands, are all parts of the rim of a saucer-shaped structure which largely determines also the outlines of Lakes Michigan and Huron. In fact, the four southern Great Lakes, Michigan, Huron, Erie, and Ontario, all belong in the same general category. They all lie, as the geologist would say, "in lowland belts eroded out between the cuestas of an ancient coastal plain."

The explanation for Lake Superior, and therefore for Keweenaw Point, is a different matter. Keweenaw Peninsula is not in the same category with the other peninsulas just mentioned. Nor is Isle Royale comparable with the Manitoulin Islands.

Keweenaw Point was at one time the greatest copper-producing locality in the United States, although it is now exceeded by the greater mines of Utah and Arizona. Isle Royale, which is about fifty miles long, the same length as Keweenaw Peninsula, is the largest island in Lake Superior. It is the wildest national park in the United States, not having even a road. Abounding in moose and other native animals, and supporting a splendid virgin forest, it is a spot of unique beauty. It is probably less accessible to visitors than any other national park. Because Isle Royale and Keweenaw Point are closely related in origin, attention has been directed to both of them in this example.

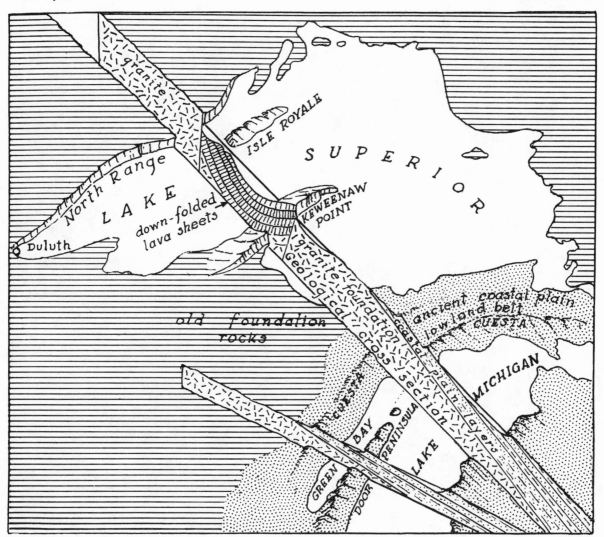

An examination of the above diagram shows the similarities and the differences between Keweenaw Point and other peninsulas in the Great Lakes region, such as the Door Peninsula of Wisconsin. We note first of all that Lake Superior occupies a long depression formed by the down-warping of part of the earth's crust. It is, geologically speaking, a synclinal basin or trough. Part of this trough consists of old lava formations or trap sheets, the layers represented by the darker pattern on the above diagram. The south side or "limb" of this down-warped structure forms Keweenaw Point. The north side, where the fold comes up again, emerges above the level of Lake Superior to form Isle Royale. Isle Royale is simply the counterpart of Keweenaw Peninsula.

Glance now at the lower part of the map and note the structure of Lake Michigan and the Door Peninsula. Here the formations dip off gently to the southeast. Certain beds here are more resistant to erosion than others, and they project to form such ridges as the Door Peninsula. The water bodies, such as Green Bay and Lake Michigan, lie in the lowlands that have been eroded out on the weaker beds between the ridges. These dipping beds form a relatively thin layer above the granite foundation underneath. They represent the layers of an ancient coastal plain, and at one time were deposited under the sea. The beds forming Keweenaw Point, on the other hand, are much older formations, for they are part of the old foundation upon which the coastal plain beds rest.

EXAMPLE 11   *The Problem*

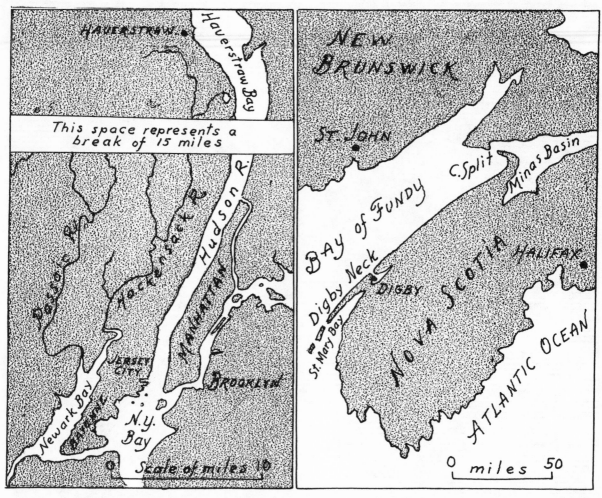

SHORE LINES. *Two Similar Peninsulas. Bayonne Peninsula in New York Harbor; Digby Neck, Nova Scotia.*

The Port of New York, with its more than 500 miles of waterfront, is one of the most commodious and remarkable harbors in the world. Into New York Bay converge the three great estuaries which form the Hudson River, the East River, and Newark Bay. Although Newark Bay happens to be in New Jersey, it is, nevertheless, part of the New York waterfront. Between Newark Bay and New York Bay the Bayonne Peninsula projects south from Jersey City. On both sides of this peninsula are numerous wharves and docks. Here come many of the foreign freighters, and from here, too, go the barges which bring to Manhattan Island the vast quantity of produce that arrives here by rail from all parts of the United States.

The peninsula of Digby Neck in Nova Scotia contrasts with the Bayonne Peninsula in being wild and almost undeveloped. There are no large cities or ports nearby. Nevertheless, from the standpoint of its origin, it is an almost duplicate of the Bayonne Peninsula. This being the case, we would expect to find also that many other features in the Nova Scotia region have their counterpart in the New York City area, perhaps on a smaller scale. The two areas in fact are almost perfect images of each other in many of their details. The two maps above present the salient geographical facts, but they hardly provide a clue to the real meaning of the several features there presented. It is only when we analyze the two areas and explain them in terms of their geological structure that we are able to discern some of the interesting relationships which the maps above do not tell us.

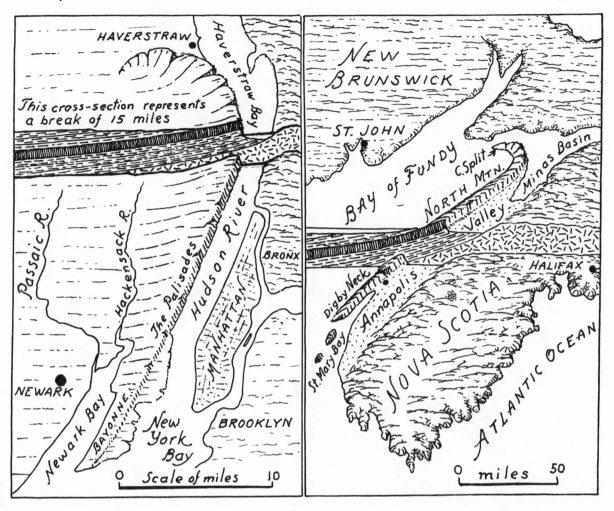

The above illustration clearly shows that the Bayonne Peninsula is a continuation of the Palisades. The cross section shows the structure of the Palisades. This structure is relatively simple. A layer of dark igneous rock, a former molten lava sheet, lies between other layers of shale. The dark igneous rock is tougher and more resistant than the shale, and therefore forms the high ridge of the Palisades. The shale beds *under*lying the igneous layer have been worn down along their outcrop, and the lowland thus formed is occupied by the Hudson River and its enlarged northern portion that is called Haverstraw Bay. West of the Palisades the broad lowland formed on the *over*lying shales constitutes the Hackensack and Passaic valleys. These together are actually a part of the so-called Triassic Lowland of northern New Jersey. The Palisades at their northern end swing around in a circular arc at Haverstraw because here all the beds have been warped to form a saucer-shaped basin.

Coming now to the Nova Scotia region, let us note some of the details which have their counterparts in the New York area: (*a*) Digby Neck corresponds with the Bayonne Peninsula. It is the southern end of North Mountain, which is an almost exact duplicate of the Palisades. (*b*) The curving north end of North Mountain at Cape Split is similar to the curving north end of the Palisades, and for the same reason. (*c*) Minas Basin corresponds with Haverstraw Bay. (*d*) The Annapolis Valley in Nova Scotia corresponds with the trench of the Hudson River. (*e*) The Bay of Fundy corresponds with the Hackensack Valley. (*f*) Nova Scotia itself corresponds with the rugged land of Westchester County, the Bronx, and Manhattan Island.

EXAMPLE 12   *The Problem*

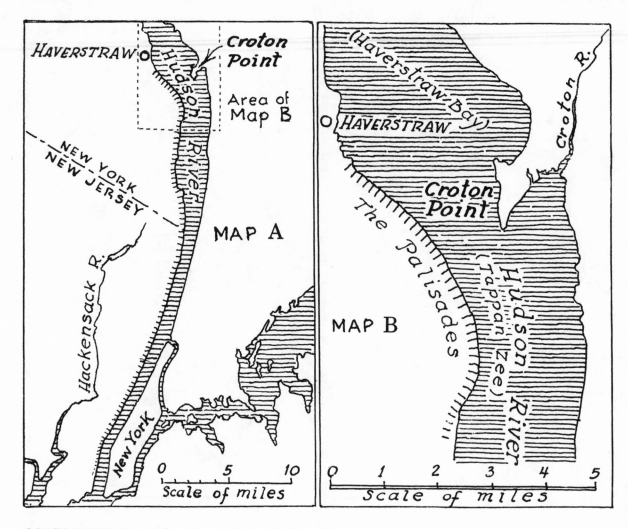

COAST LINES. *Peninsulas. Croton Point, Hudson River.*

Some 25 to 30 miles north of New York City, the Hudson River (which is actually not a flowing river but an estuary or arm of the ocean) opens out into a large bay or inland sea. This body of water, some five miles across, is called Haverstraw Bay or Tappan Zee. Projecting out into this bay from its eastern side, just north of the city of Ossining, is a peninsula reaching halfway across the river. It extends outward some 2½ miles. The left-hand map above shows the location of this point with relation to New York City. The map on the right is considerably enlarged in order to show the actual shape of this point. Being near the mouth of the Croton River, it is called Croton Point. These two maps, except for the omission of the roads and towns, show the details just about as they appear on some of the automobile road maps, and also on about the same scale. No information is available on maps such as these as to the character of this Point, whether it is a rocky promontory or a flat plain. The road map indicates that it is a State Park, but this fact does not convey any information as to how it got there in the first place. It is certainly not an artificial feature like the long pier which extends out into the river at Piermont. It is indeed a rather unusual feature. There are no other features like this anywhere else along the Hudson River between New York and Albany. As we seek an explanation for this unique feature we may be reminded that we have here a problem like that presented in many mystery stories. The clue is actually somewhere on the map itself.

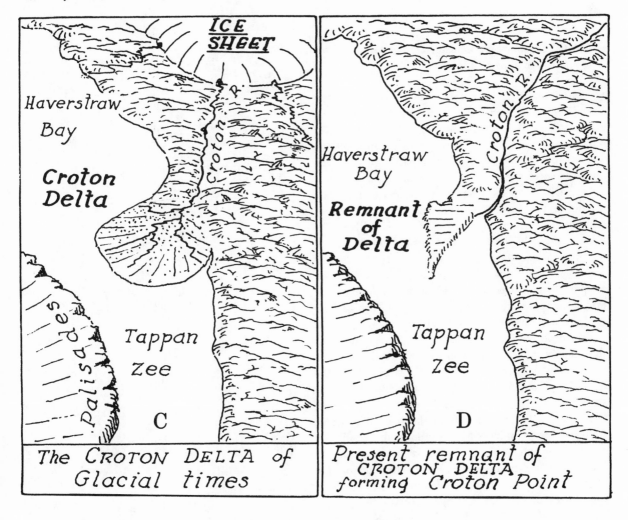

The CROTON DELTA of Glacial times

Present remnant of CROTON DELTA forming Croton Point

The clue to the origin of Croton Point, to which we just referred on the previous page, is the Croton River. During the waning stages of Glacial Time, when this part of the continent stood somewhat lower than it does now, because of the great weight of ice upon it, the Hudson River was about 80 feet deeper in Haverstraw Bay than it is at present. The Croton River, pouring out from the melting ice front, carried great quantities of sand and gravel into Haverstraw Bay and built there a large delta which reached halfway across the river. Like most deltas built into quiet estuaries, the Croton Delta was more or less round in shape, with distributary streams flowing outward in all directions toward its margins.

Following the final disappearance of the ice and the removal of this great weight, the crust of the earth in this part of the United States gradually rose above sea level. In the Croton Delta region the elevation was about 80 feet, with the result that the flat top of Croton Point stands now 80 feet above sea level. An important result of this rising was the invigorating effect it had upon the Croton River. This stream, therefore, flowed more swiftly, and eroded its valley extensively. Much of the delta was removed by the river, so that now only the northern half remains. This is clearly revealed by its present shape.

Although no other deltas occur along the Hudson River, nevertheless the old shore lines of the river are to be seen near Albany, some 200 feet above the present river level. Farther north, in Canada, the rising of the continent since Glacial Time was even greater.

EXAMPLE 13

The Problem

COAST LINES. *Some Coastal Irregularities. The Hudson River Shore Line.*

Reaching inland for 150 miles north from New York City is a long narrow arm of the sea known as the Hudson River. It is not at the present time a real river but a drowned estuary. The tide actually runs all the way up to Troy. On the preceding pages we have just noted some of the characteristics and features of this estuary, with particular reference to the remarkable delta of the Croton River. Our concern here is to explain some of the smaller irregularities of its shore line. The maps at the right show a portion of the Hudson estuary for some distance north of New York City. The larger-scaled one, the one on the left-hand side, depicts the scarp of the Palisades for a length of three miles or more just north of the great George Washington Bridge. The exact location of this area with relation to New York is indicated on the Index Map at the right. In particular, the features to be noted along the shore line are the small crenulate irregularities along the river edge at the base of the Palisades. The Index Map shows that these small, almost invisible irregularities are lacking along the shore on the eastern side of the river. The Index Map has been reproduced on a scale made familiar by many of the automobile road maps, a scale too small to reveal clearly the minor details. For that reason the larger scale map has been included here. It is a redrawing of part of the U.S. Geological Survey topographic sheet, and slightly reduced. Some smaller details have been added. The Palisades are represented by hachures as a steep cliff. The upper half, that is, about the uppermost 200 feet of it, is almost vertical. The lower half slopes off more gently to the river. A road, known as the Henry Hudson Drive, is represented as a single broken line running the full length of the map, and for much of the way at the break in the slope halfway between the top and bottom of the Palisades scarp.

The entire shore line, together with the scarp of the Palisades, shown on the larger-scale map is included in the Palisades Interstate Park, which embraces an even greater area farther north. Most of the region is therefore available to pedestrians, especially to those who follow the trail along the river's edge, and much of it to those who travel by car.

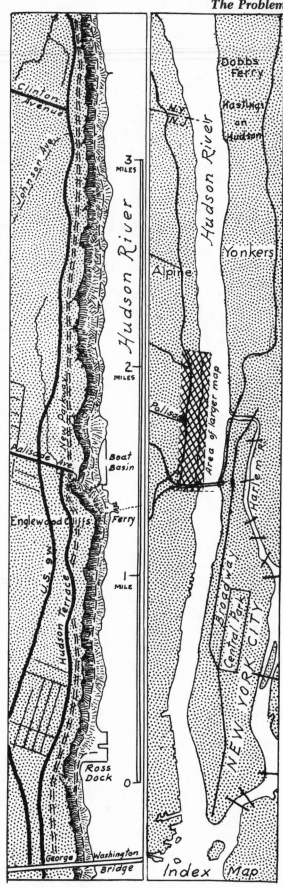

Index Map

columns

River

The sketch above illustrates one of the small irregularities in the shore line at the base of the Palisades. This irregularity is the result of a landslide. In the landslide pictured here, a mass of rock has broken off from the face of the cliff and has left a white scar there. The broken fragments of rock have torn their way down through the forest and have obliterated many of the trees. At the river's edge this mass of debris spread out in a fan-like form and pushed its way into the water. Many such landslides have taken place along the front of the Palisades in the past, and have since been covered with a growth of trees.

Some landslides do not occur suddenly and abruptly. Instead, blocks of rock drop off the cliff from time to time and accumulate below as a long talus slope. This breaking off occurs usually during the late winter and spring when the ice in the crevices of the rock starts to thaw. This causes the rocks to give way. In the picture above, a small rock slide is shown to the left of the larger one. In this case a relatively small mass of rock broke off and tore a narrow path through the forest all the way to the river's edge. In innumerable instances the fragments of rock simply roll down among the trees, doing no appreciable damage, but building up gradually the large mass of talus accumulations.

The picture above was made near Alpine, which is just north of the limits of the large-scale map on the opposite page. The Henry Hudson Drive ascends the Palisades at Alpine and therefore does not appear in the picture. It has already been noted that the Drive has been constructed along the upper part of the talus slopes and just at the base of the steep cliffs. This means therefore that rocks breaking off the cliff first fall on the roadway, where they are a hazard to traffic. For this reason the Drive is closed to the public during the winter and early spring months.

The rock falls of the Palisades are petty things compared with the avalanches and great landslides of some mountain regions. In many instances entire valleys have been blocked up by debris so that lakes have resulted, and even roads and villages have been wiped out.

EXAMPLE 14   *The Problem*

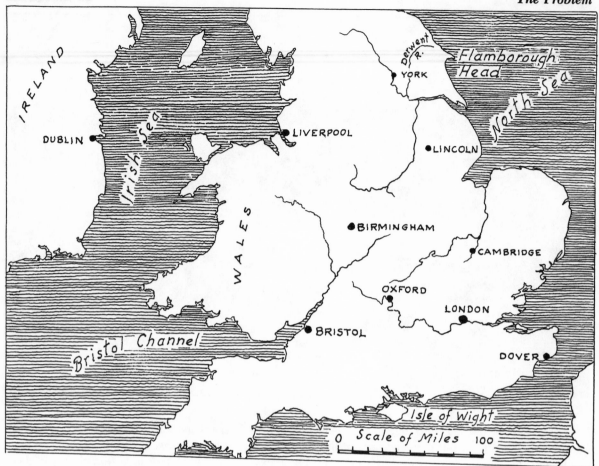

COAST LINES. *Promontories and Points. Flamborough Head, England.*

The next point or promontory we are going to examine is along the eastern coast of England. It is known as Flamborough Head. Its small size suggests that it is a spit of some kind, but it is rather too blunt for this. On a map of England, Flamborough Head is almost too small to notice. Nevertheless it is actually part of the larger pattern of southeastern England. It is related even to the Isle of Wight on the English Channel.

This headland has its counterpart in many other parts of the world. A couple of similar examples have already been described from the United States, in Example No. 9.

When we discover the true explanation for Flamborough Head, we shall see that this explanation also explains many of the features of southeastern England. The two Ouse rivers, the Trent, the Avon, and even part of the Thames River fall into this larger pattern. So also do many of the cities such as York, Lincoln, Cambridge, and Oxford. As you examine Flamborough Head notice also the Derwent River, how it rises near the coast and flows westward in a curve which is the continuation of the curve which forms Flamborough Head.

The coast line of the British Isles has many irregularities. But we notice that the coast of southeastern England is much simpler and smoother than is the coast line of Wales, Ireland, and western England. There, in the west, we notice many small estuaries and promontories of all shapes; very irregular shapes they are. None of them have the prong-like form like that of Flamborough Head. It is clear, therefore, that the western half of the area shown on the above map is quite different from the eastern half of the area. Not only is the coast line different but the topography of the interior part of the country is also different. Thus we discover that the insignificant details of the coast line of a country provide a clue to the larger features of the country itself.

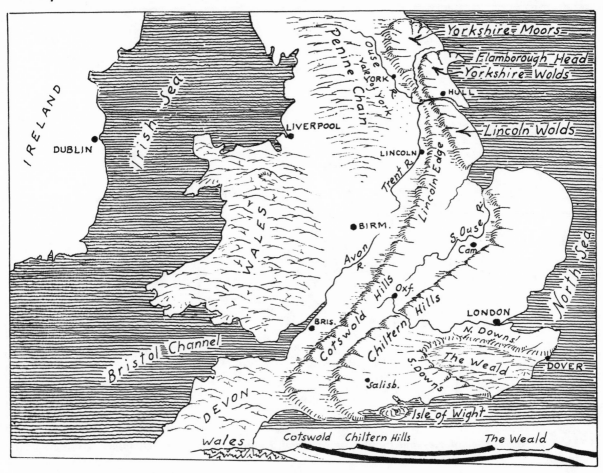

Flamborough Head, like Door Peninsula and Saugeen Peninsula already described from the Great Lakes area in the United States, in Example No. 9, is the tip end of what is called a cuesta. It is not a sand spit, but is part of the rocky framework of England. Like a protuberance of a bony skeleton, it seems to be forcing its way out of the skin.

The diagrammatic map above shows that this point is the northern terminus of the long curving scarp, or cuesta, whose southern end is represented by the Isle of Wight. Its middle portion is known as the Chiltern Hills, a little way northwest of London, and its northern portion forms the Lincoln Wolds and the Yorkshire Wolds.

A second and almost parallel cuesta forms the Cotswold Hills, which farther north is known as the Lincoln Edge and the Yorkshire Moors. The Yorkshire Moors do not quite stick out to form a promontory like Flamborough Head. The southern end of this winding scarp forms the peninsula of Purbeck Downs on the English Channel.

The cross section below the map shows the two hard layers of rock which form these two cuesta scarps. The hard layers are separated from each other and also from the underlying complex basement of older rocks by weaker formations of clays. These clay beds underlie the two lowlands which form long belts just west of the two scarps. Draining the two lowlands are the several rivers that have been mentioned; namely, the North Ouse, the Trent, and the Avon in the western lowland; and the Derwent, the Southern Ouse, and the upper Thames in the eastern lowland. This eastern lowland, because it contains the cities of Oxford and Cambridge and other famous schools, is known as the "Educational Lowland" of England. The rock layers forming the scarps are limestone beds. The upper limestone bed which forms the Chiltern Hills is a chalk formation which comes out again in the North Downs and the South Downs surrounding the Weald Dome. It also forms the "White Cliffs of Dover."

29

EXAMPLE 15 *The Problem*

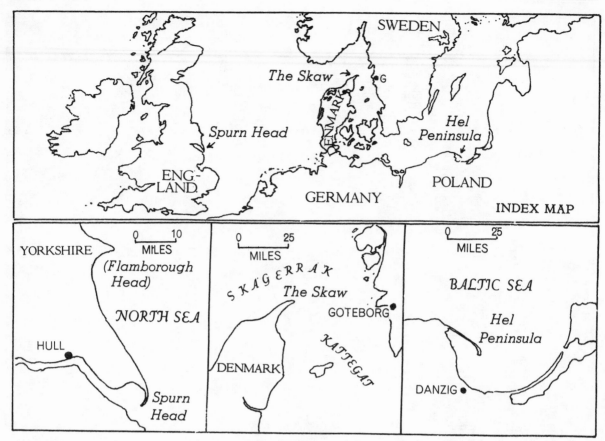

COAST LINES. *Promontories and Points. Spurn Head, England; The Skaw, Denmark; Hel Peninsula, Poland.*

At many places around the North Sea and the Baltic are points of land that project outward from the coast. They do not resemble the irregular type of promontories that occur in such profusion along the coast of England and Wales and elsewhere throughout the world. They are smaller, much sharper, and more pointed.

Note first the tip end of Denmark. This long point is known as Cape Skagen or The Skaw. It projects toward the Swedish coast for ten miles or more. There are many other similar but smaller points along the coast of Denmark.

Farther east, in Poland, along the southern coast of the Baltic Sea, is an unusual projection which reaches far out into the Gulf of Danzig. This point goes by the name of the Putziger Nehrung, or simply the Hel Peninsula. It is fully 20 miles long and barely ⅛ of a mile across throughout most of its length.

Elsewhere in the world are scores of other promontories and points. Spurn Head on the North Sea coast of England is a noteworthy case. Flamborough Head, a few miles farther north, though somewhat resembling Spurn Head, is a different kind of land form, with quite a different origin. Cape Farewell, at the north end of the South Island of New Zealand, may also be noted.

All of the examples cited above, The Skaw, Spurn Head, and Hel Peninsula, are actually unusually large features of their type. Most geological features formed in the way these have been are relatively small, in most instances not detectable on small-scale maps like those in atlases, and those which show entire continents. It behooves us in examining such maps not to assume that all small points and projections belong in the same category. The eastern tip of Long Island, the southern end of Florida at Key West, and the eastern end of Cyprus at Cape Andreas all look more or less alike, but each one has had an origin different from the others and different also from that of The Skaw and Spurn Head.

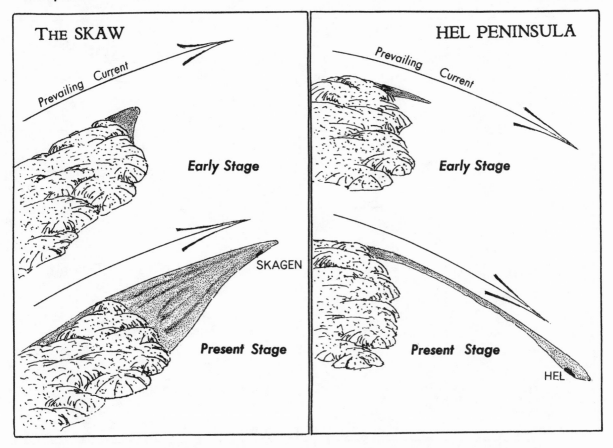

THE SKAW

Prevailing Current

Early Stage

Present Stage

SKAGEN

HEL PENINSULA

Prevailing Current

Early Stage

Present Stage

HEL

Geological features like Spurn Head, The Skaw, and Hel Peninsula have been formed very recently and are still growing. They are known as "spits," a spit being a point, like the spit or prong upon which a roast is turned over a fire. A spit is a point of sand which has advanced outward from a headland. The sand is derived from the wearing down of the headland by the waves. It is drifted outward by the prevailing currents, being constantly fed by more and more sand that is carried alongshore. That this actually takes place is attested to by the fact that the sand grains constantly become finer, the farther they are carried away from their source.

Stages in the development of both The Skaw and the Hel Peninsula are portrayed above. All spits are formed in this manner, some long, some short, some broad, some narrow. Many, because of the blowing winds, are surmounted by sand dunes. Others become clothed with vegetation as the dunes gradually become fixed. Most spits have splendid beaches which may be either sandy or gravelly, and subject also to frequent change in character with the changing currents, waves and wind.

The Skaw has been built out from a promontory some 150 feet high, consisting of irregular hills of readily eroded glacial deposits. Its broad surface is ribbed with several beach ridges which indicate successive stages in its growth. Its surface, too, is covered with numerous sand dunes some of which attain heights of 100 feet. Skagen, the port at its far end, stands on the beach, its harbor being protected by an artificial breakwater.

The Hel Peninsula, like The Skaw, juts out from a headland of high hills. But, unlike The Skaw, it is extremely narrow. Its far end, however, swells out to more than a mile in width. Here is the port of Hel, situated, like Skagen, on the inner, protected side.

From this it can be seen that eastern Long Island with its morainal hills, southern Florida with its coral islands, and Cyprus with its long rocky headland all come under categories totally different from the spits illustrated above.

EXAMPLE 16

The Problem

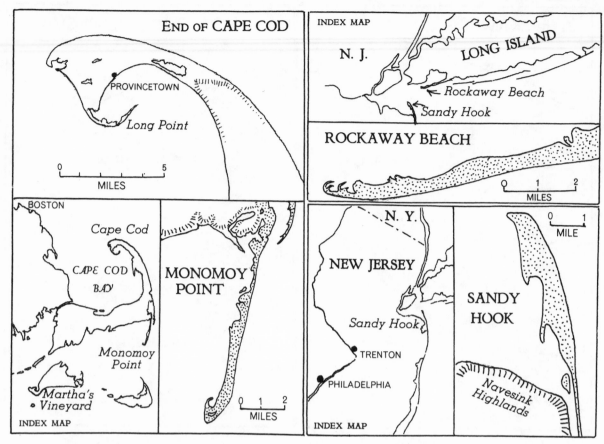

COAST LINES. *Promontories and Points. Hooked Points. Cape Cod; Monomoy Point; Sandy Hook; Rockaway Point.*

Many of the points found on maps are not so simple as the ones described in the last example. Instead of coming to a sharp point they end in hooks. And not only that, but most of them have barbs or smaller projections jutting out from their inner or protected side, the side away from the open ocean. Four striking examples are shown above, just as they appear on readily available maps of moderate scale. They are all well known to literally millions of people. Less well known examples occur elsewhere throughout the world. Let us examine each one of these for a moment and see what features they have in common. These features can be enumerated as follows:

1. All of these points curve back at their ends to produce hooks. Some of the hooks are almost complete circles, others have shorter arcs.

2. There appears to be at the end of the hook, and on its inner side, an additional little hook, as if it were an afterthought. This additional hook is in some cases big enough to have its own name, as for example Long Point at the end of Cape Cod. This point forms a natural breakwater for the port of Provincetown.

3. The barbs, mentioned in the first paragraph above, are particularly noteworthy on Rockaway Beach and Sandy Hook, and also on the small unnamed spit north of Monomoy Beach.

Each of these points, or beaches, is 5 to 10 miles in length, and conspicuous enough therefore to be shown on most maps. The profuse detail with which they are embroidered is, however, not shown on these small scales. For these details it is necessary to consult the topographic maps of the U.S. Geological Survey or the various charts of the Coast and Geodetic Survey. As most of these maps have been made from aerial photographs, the details which they portray are surprisingly accurate. In many cases the underwater features, as revealed on the coast charts, help us to understand how these complicated land forms have come about.

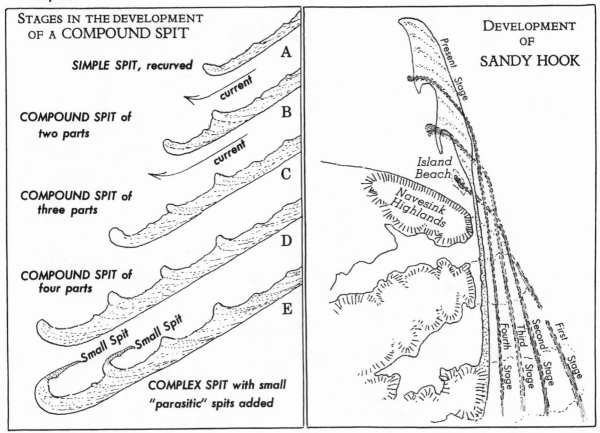

STAGES IN THE DEVELOPMENT OF A COMPOUND SPIT

SIMPLE SPIT, recurved — A

COMPOUND SPIT of two parts — B

COMPOUND SPIT of three parts — C

COMPOUND SPIT of four parts — D

COMPLEX SPIT with small "parasitic" spits added — E

DEVELOPMENT OF SANDY HOOK

Some of the steps in the development of hooked points are shown above, in the sketches at the left. Starting off with a simple spit, we note that there is a tendency for it to turn back at its tip. This occurs when the current which produces it is too weak to maintain a continuous straight line, but instead swings into any embayment which happens to be present. With the shallowing of the water by the growth of the spit, the current gradually resumes its straighter course until it again turns inward to form a second barb. The spit now, as in "B," has become a "compound spit" because it is really made up of two spits fused together, or compounded. Similarly, in "C" there are three fused spits, and in "D" there are four of them. In the final stage, shown in "E," the ends of some of the barbs have been attacked by smaller waves and currents, with the result that some very small baby spits have been added, almost like parasites upon the larger one. All of this is a fairly close approximation to the history of Rockaway Point, as well as of Monomoy Point.

In the illustration on the right, above, the several steps in the growth of Sandy Hook are shown. It is to be noted that in the earlier stages the land area of New Jersey extended farther into the Atlantic Ocean. As the land was worn back the spit gradually assumed its present position. Island Beach represents the mere remnant of the second stage.

In recent times the additional small hooks or secondary spits have been added on the inside of the main hook.

Land forms, such as beaches and spits, pass through their stages of development with relative rapidity. Changes can be noted even within a person's lifetime. Therefore, these forms have served to emphasize the idea of evolution of land forms of all kinds. Mountains, rivers, even plains and plateaus also exhibit stages of development. By studying large numbers of these forms throughout the world, it is possible to recognize what the sequences are, even though from year to year the actual changes are imperceptible.

EXAMPLE 17   *The Problem*

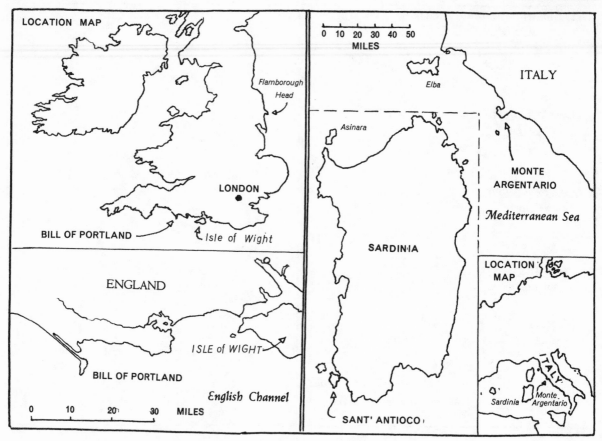

COAST LINES. *Promontories. Bill of Portland, England; Monte Argentario, Italy.*

Most coast lines have many small irregularities which seem to be of little significance, but which when examined on larger-scale maps are found to have peculiar interest. Such, for example, are the ones to which attention is directed on the above maps. It is of course true that atlas maps rarely provide a clue as to how these features have come about; nevertheless they encourage us to wonder about such matters.

On the coast of southern England, along the English Channel, is a minute projection which goes by the name of "Bill of Portland." On small-scale maps it is the merest point. On larger maps, as on the lower one above, it is seen to have a somewhat more definite shape and character. And on still larger maps, like those of the British Ordnance Survey, there is sufficient detail to provide a real clue to its origin. This point is virtually the only one of its kind along this part of the English coast, or for that matter along the entire coast of England or Wales. Nevertheless, there are many other examples of its kind throughout the world.

Another example of this kind of promontory, but somewhat more involved in its details, is the projection along the west coast of Italy which is known as Monte Argentario. Still another example occurs near the southern end of the island of Sardinia. It is known as Sant' Antioco. The chief characteristic of all of these points is the narrow isthmus which connects it with the mainland. With this in mind it is interesting to explore our atlas for still more examples. Perhaps in doing so we shall run across the southern tip of Australia, just north of Tasmania. This point is called "Wilson's Promontory." It might be expected that we would have little trouble finding other examples. The fact is, however, that features of this kind are usually so small as to defy detection, except on large-scale maps. If they are big enough to show clearly on atlas maps, as for example the peninsula at the Cape of Good Hope, they are probably features of a different sort.

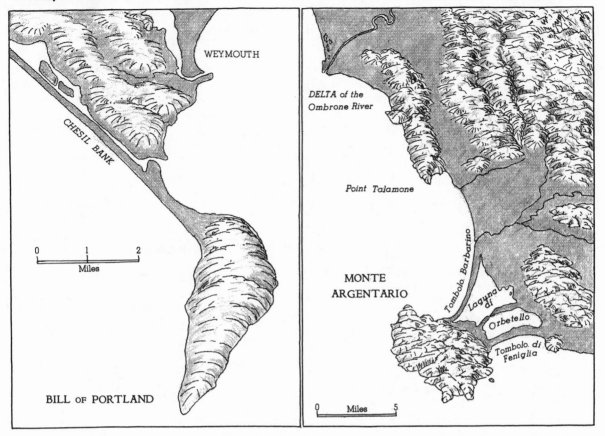

The type of feature to which attention has just been called, such as the Bill of Portland and Monte Argentario, are known geologically as "tombolos," or land-tied islands. At one time in the not so distant geological past (measured probably in thousands of years) they were islands lying offshore. They have become land-tied by the building of sand bars that have attached them to the mainland. The Bill of Portland is a hill some 300 feet high. As shown on the above map, it is several miles long and tapers off toward the south. This former island lay at one time about 2 miles offshore. The bar of sand which now joins it to the land is, however, 10 miles long, much too long to be shown on our map. This bar is now covered with sand dunes, and is known as Chesil Bank.

Monte Argentario, on the Italian coast, is perhaps even more unusual and interesting. This former island, virtually an outlier of the Apennines, is much larger than the Bill of Portland, for it is a good 5 miles in diameter and rises over 2,000 feet above the sea. It is a veritable mountain. Three bars, called "tombolos" by the Italians, now join this mountain to the mainland. These bars are called tombolos because of the "tumuli" or sand hills which surmount them. But the term "tombolo" has come to mean also the entire land-tied island.

On the previous page it was suggested that possibly Sant' Antioco on Sardinia is also a feature of this sort. It does appear so on most maps, but as an actual fact it is not quite connected with the mainland, although it is joined by a railroad.

It is interesting to surmise that perhaps the island of Elba, shown on the map on the opposite page, will some day also become a tombolo.

EXAMPLE 18 *The Problem*

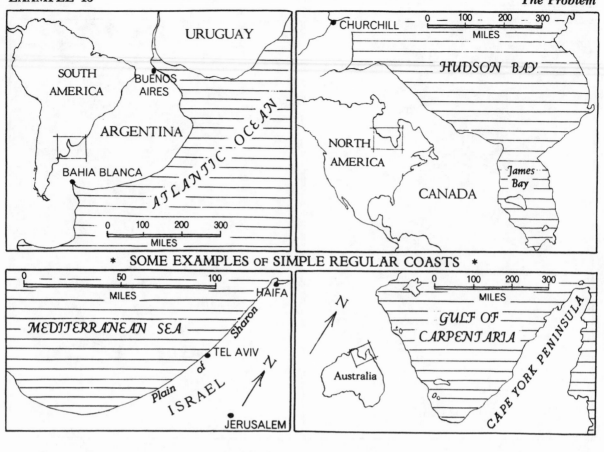

* SOME EXAMPLES OF SIMPLE REGULAR COASTS *

COAST LINES. *Simple, Regular coasts. Argentina; Hudson Bay; Israel; Gulf of Carpentaria, Australia.*

Most coast lines appear to be irregular. Our problem in this example is not to explain some of the irregularities of coasts, as in the previous examples, but to explain why some coasts are not irregular. The normal type of coast, it would appear from the inspection of many maps, is the irregular one; the unusual kind is simple, smooth, and uninterrupted. Those of us who live in the eastern United States are familiar with both irregular and simple shore lines—the irregular shore lines of Maine on the one hand, and the simple uninterrupted beaches of Long Island and New Jersey on the other. The beaches of Long Island and New Jersey are, however, merely long bars of sand. They are not, strictly speaking, the actual edge of the mainland, which is itself somewhat irregular.

The above map reproduces from four different continents some striking examples of long, simple uninterrupted coast lines. The coast of Argentina in South America makes a long sweeping curve from Buenos Aires southward to Bahía Blanca. Throughout this whole stretch, for hundreds of miles, there is no harbor nor any significant port. In fact there is hardly a town anywhere along this coast. It is low-lying, and shallow water extends far offshore.

A less familiar coast is the vast stretch along the western side of Hudson Bay, in Canada. Here, again, the water is remarkably shallow. Changing winds are sufficient to change the position of the shore line a mile or more, and of course the tides are even more effective than the wind in altering the position of land and sea. In out-of-the-way places of the world like this, a simple uninterrupted shore line on the map may merely mean that the coast line has not been explored in detail. In such places crude maps may be drawn from the decks of passing vessels unable to come close to land. However, with present-day mapping methods based upon aerial photography, few coast lines of the world are now unknown in detail.

The Plain of Sharon in Israel presents toward the Mediterranean a simple almost straight sweep of low coast line; and in Australia the Gulf of Carpentaria, like Hudson Bay another out-of-the-way part of the world, has also a low regular coast line devoid of any bays or promontories.

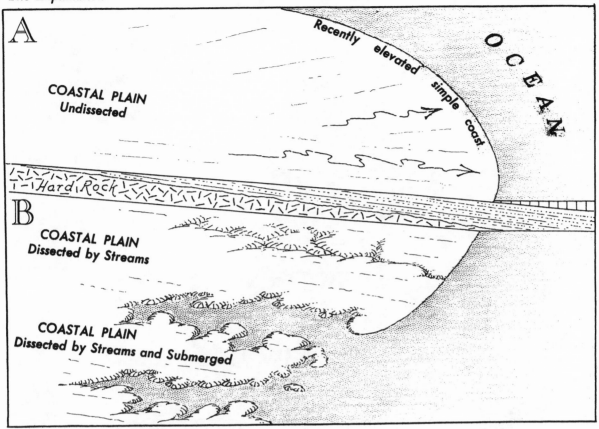

The simplest type of shore line is that which is found along the margin of a coastal plain. A coastal plain is an upraised or upwarped part of the old sea floor. It is made up of loose, unconsolidated sediments, usually clay, sand, and marl originally deposited under water. Such a sea bottom is apt to be fairly flat because original depressions become filled with the accumulating sediments, and the constant movement, alongshore currents, and the like, tend to smooth off any unevenness of the deposits. A relatively flat sea bottom formed in this manner, when raised above sea level, becomes a "coastal plain" (Figure A above). Its margin where it comes in contact with the sea is simple and regular, sometimes practically straight. When the elevation of a coastal plain above sea level occurs, it is often accompanied by a certain amount of warping and bending. This means that the new coast line, instead of being straight, will be sinuous, curving around the bowed-up portions and swinging inland into the more downwarped parts.

As time goes on, a coastal plain becomes eroded by streams. These form shallow valleys, usually with many tributaries coming in from the sides (Figure B). This erosion does not affect the shape of the shore line. However, a coastal plain is rarely stable for long. "It rises or falls as glaciers melt or grow, as the floor of the deep ocean basins shifts under its increasing load of sediments, or as the earth's crust along the continental margins warps up or down in a continuing adjustment to strain and tension. . . . Always the edge of the sea remains an elusive and indefinable boundary" (Rachel Carson in the *New Yorker*).

When the sea level rises as a result of these changes, the previously formed stream valleys become drowned or submerged, and the shore line therefore becomes irregular. This is shown in the foreground of the illustration above. Simple coasts, as a consequence, are merely temporary conditions, the trend being always toward the formation of irregularities.

EXAMPLE 19   *The Problem*

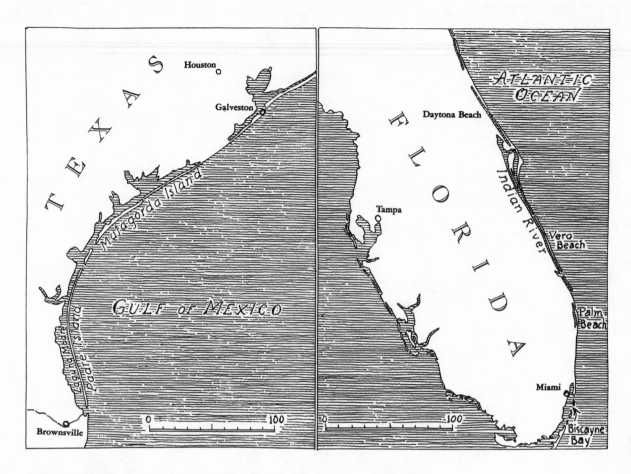

COAST LINES. *Coast Lines with Barrier Islands. Texas and Florida.*

An examination of almost any map of Texas and of Florida reveals the remarkable long offshore bars that form the coast line. Between these bars and the mainland is a long narrow lagoon. The southern member of the series of bars in Texas is called Padre Island. It is more than 100 miles long. The lagoon back of this bar is called Laguna Madre. Farther north along the Texas shore line is another long bar known as Matagorda Island. There are other, similar bars, some of which touch the mainland to form peninsulas. The city of Galveston stands on a bar of this kind. It is reached by rail and by road only over long causeways.

Along the coast of Florida are similar long offshore islands which here and there touch the mainland. Daytona Beach, Vero Beach, and Miami Beach are all well known examples. Behind each of these beaches is a lagoon similar to Laguna Madre. The Indian River portion of the lagoon is more than 100 miles long. It serves as part of the long Intracoastal Waterway which provides a continuous protected water route for small craft all the way from New York to Miami. The Indian River region, too, is well known for its extensive orange groves. In the Miami region, the lagoon corresponding with Laguna Madre and with Indian River is called Biscayne Bay.

The entire Atlantic coast south of New York is characterized by offshore bars like those of Florida and Texas. Atlantic City and other resort towns are situated on these narrow islands of sand separated from the mainland by bodies of water, such as Barnegat Bay in New Jersey. Elsewhere in the world corresponding forms may be observed along the Baltic, the coast of India, and here and there along the coast of Africa and of Australia.

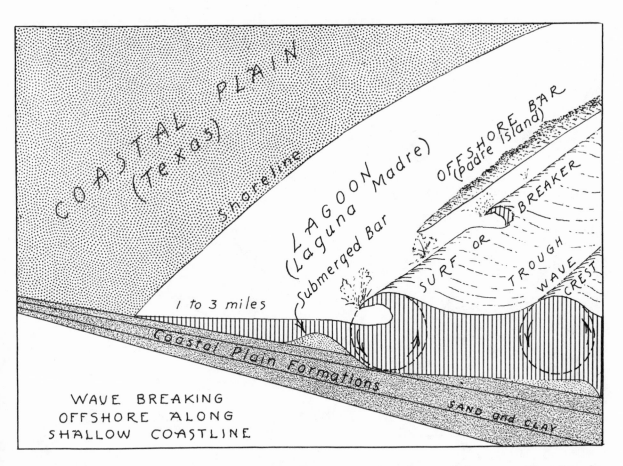

Offshore or barrier bars form only along coasts which shelve off gently into shallow water. Rugged, rocky coasts never have features of this kind. In other words, barrier bars occur along the margins of coastal plains, or along the margin of deltas and other similar low-lying areas.

The great Atlantic and Gulf Coastal Plain which extends from Cape Cod to Mexico exhibits offshore bars throughout its whole extent. The explanation for this is given in the diagram above. In the foreground of this illustration we see a geological cross section of the Coastal Plain. The Coastal Plain formations dip off gently from the low sloping mainland of Texas beneath the shallow waters of the Gulf, just as they do also along the Atlantic shore. We note also the waves rolling in toward the land where they break offshore. Because of the shallowness of the water the waves can not reach the actual shore line. The wave motion of the water resembles a big cylinder which may extend far beneath the surface of the water, even 50 feet or more, depending upon the violence of the wind which created the wave. As a result of this, the wave scrapes on the bottom far offshore, long before it reaches the mainland. The sand that is eroded is thrown up in front of the wave as it breaks so as to form a bar. In its early stages the bar is low and is submerged. But as it increases in height it forms an offshore island. Between the island and the mainland is the so-called lagoon. This narrow body of water varies in width up to 3 miles or more. The lagoon soon becomes filled with marsh grass and with silt. In time it may completely disappear or be represented merely by a winding bayou or channel way.

As the offshore bar increases in height and in width, it becomes covered with sand dunes. Between the dunes, at times of severe storms, the surf may pour over the island and into the lagoon. The opening thus formed may become a permanent tidal inlet.

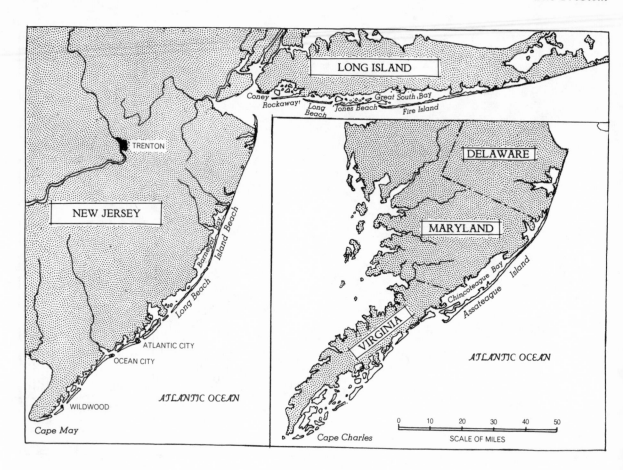

COAST LINES. *With Interrupted Barrier Islands. Long Island; New Jersey; Delaware-Maryland-Virginia Peninsula.*

The three areas depicted here are similar in so many respects as to command especial interest. Note that in each case the barrier bar at its eastern end (Long Island) or at its northern end (the New Jersey and Virginia examples) touches the mainland. Proceeding then southward, or westward, along the coast it becomes an offshore island separated from the mainland by a wide lagoon. In this intermediate portion the bar is almost continuous and unbroken. On Long Island this portion is called Fire Island, the lagoon back of it being known as Great South Bay. In New Jersey the two long island segments are known as Island Beach and Long Beach. Back of this long strip of beaches is Barnegat Bay, which is an almost exact replica of Great South Bay. Along the Delaware-Maryland-Virginia coast the long unbroken beach is known as Assateague Island. The lagoon back of it is called Chincoteague Bay.

There then follows in each of the three cases a number of much shorter bars and islands and broken-up marshlands intersected by bayous and channels. On Long Island these small segments include Jones Beach, the Rockaways, and Coney Island. In New Jersey these broken patches of beach support such resorts as Atlantic City, Ocean City, Wildwood, and Cape May. In Maryland and Virginia this ragged coast extends all the way to Cape Charles.

Not only in these and in several other details do these three strips of coast resemble each other, but they have also almost the same length—just over 100 miles. A pattern of forms repeated in this way presents a problem of much interest to the student of shore lines, the answer to which is provided on the following page.

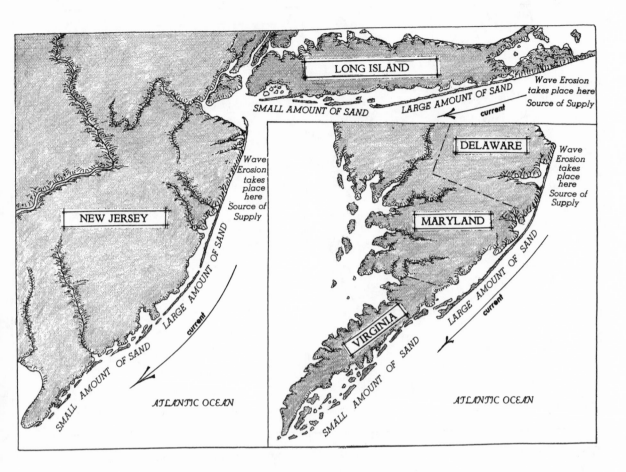

The maps above show diagrammatically how the conditions just described were brought about.

In eastern Long Island, for instance, the long offshore bar of Fire Island touches the mainland. Along this part of the coast are steep cliffs which the waves are actively eroding. In its earlier stages the offshore bar actually lay some distance away from the coast at this point. It has since been pushed landward by the great vigor of the waves. The result is that the bar no longer exists; the waves are cutting against the mainland itself. From the bluffs which the waves are eroding, the sand and silt drift westward with the alongshore currents and contribute to the building of Fire Island and Jones Beach. The amount of sand and silt constantly becomes smaller, so that in the Rockaway and Coney Island region the bars are small and fragmentary.

In New Jersey, in exactly similar manner, the waves are cutting against the high bluffs at the northern end, at Long Branch and Asbury. The destruction at these cities is so severe that extensive jetties and breakwaters have recently been constructed. The alongshore currents sweeping southward to Cape May drag great quantities of sand along the coast, the quantity constantly diminishing in the Atlantic City and Cape May region. Therefore the southernmost bars are much smaller and more broken-up than the longer ones to the north.

In the Maryland-Virginia section the southward moving currents carry the sand from the high bluffs of the Delaware coast to the Cape Charles region. By the time it has traveled that far, 100 miles or more, much of the sand has disappeared, and there is enough for only a few small beaches and islands.

A similar situation may be noted along the coast of Holland where the Frisian Islands become smaller and smaller as one goes eastward, because the North Sea currents flow in that direction.

EXAMPLE 21

The Problem

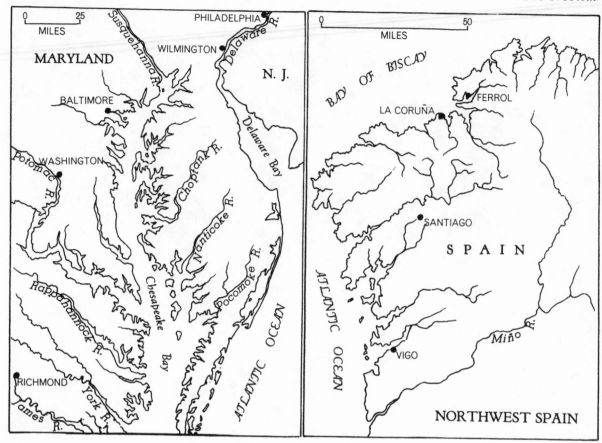

COAST LINES. *Embayed or Estuarine coasts. The Chesapeake Bay Region; the Northwest Coast of Spain.*

In each of the two examples shown above are embayments which extend far into the land. Chesapeake Bay reaches inland some 170 miles from the open sea. The embayments along the coast of Spain, however, do not reach inland so far, rarely more than 25 miles. These embayments or estuaries all occur at the mouths of rivers, both large and small. Thus the embayments in the Chesapeake Bay area have numerous branches, the larger branches having still smaller ones. The branches of the Spanish embayments, though not so numerous, are of similar character.

Chesapeake Bay is a relatively shallow body of water, and the country around it is a slightly undulating landscape, virtually a plain. It is a region of sand and clay. The embayments of the Spanish coast are deep, and the adjacent country is mountainous. Steep rocky slopes come down to the water's edge.

The Chesapeake Bay area, at the heads of the embayments, has few tidewater ports, Baltimore being the most important. Elsewhere the water is much too shallow. The Spanish coast, on the other hand, has deep-water harbors. The American ports Baltimore and Philadelphia are well developed because of their rich and easily accessible hinterland. But the Spanish ports are mostly insignificant because their location lacks the advantages which the American ports have.

A third area having exceptionally intricate embayments is to be seen on most atlases at the northern end of South Island, New Zealand.

In contrast with the embayments just described, which are fairly regular and systematic in character, are the very irregular coasts of Maine and of Greece, and the fiorded coasts of Norway, Alaska, and Chile, all of which have many large islands. Nor is their systematic branching form so clearly defined as in the examples cited above.

The problem here is not only to account for these embayments but to explain the difference in the degree to which they penetrate the land areas.

Embayed or estuarine shore lines result from the submergence of former stream valleys. The Chesapeake Bay area, for example, represents the lower course of the great Susquehanna River system, of which the Potomac is simply a tributary. These large rivers with their many smaller tributaries eroded their valleys upon the low-lying coastal plain of the Atlantic Seaboard. Only a very slight elevation of sea level was required to inundate the mouths of these valleys. In fact, the inundation extended far upstream because the valleys had such gentle slopes toward the sea. This is shown in Figure 1, above.

In contrast with the Chesapeake Bay region, northwest Spain is rugged. The embayments, or *rías*, as they are called there, do not under such conditions extend so far inland, as shown in Figure 2.

The illustrations above are designed to indicate the contrast between the low-lying and the more rugged types of coast. The left-hand side of each of the sketches represents the region before inundation. In the upper illustration the coastal plain is dissected by a stream system into a gently rolling landscape. In the lower illustration the valleys are deep and gorge-like, and the country is therefore mountainous. The steep slope, or gradient, of the valley floors prevents the sea from reaching very far inland, in case the sea rises or the land is depressed.

Submergence results from various causes. For one thing, continental masses are rarely perfectly stable over geological periods of time. The crust of the earth is always rising or falling. But perhaps the most common cause of submergence throughout the world today is the fact that the melting of the great ice sheets during the past 20,000 years or so has contributed to the oceans enough water to raise their level more than 200 feet, on the average, throughout the world.

In this connection it is interesting to note that there is enough water impounded in the present-day ice-caps and glaciers to cause, if it should all become melted and returned to the oceans, a rise of sea level sufficient to inundate most of the great seaports of the world.

EXAMPLE 22

The Problem

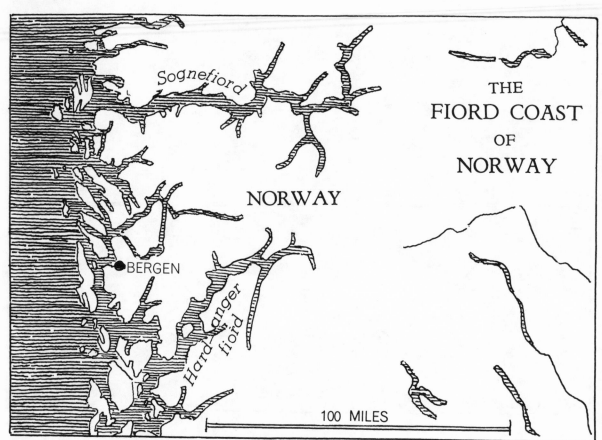

COAST LINES. *Fiord Coasts. Norway, Alaska, Chile, New Zealand.*

Differing somewhat from the type of irregular coast just described is the fiord type of coast, which is found only in four or five widely separated parts of the world. These fiord regions are as follows: Scandinavia, with its extension into Scotland; southern Alaska, and Greenland—all in the Northern Hemisphere; and Chile and New Zealand in the Southern Hemisphere. Along all these coasts the fiords or drowned valleys penetrate far inland, and are relatively narrow throughout their length. Fiords, moreover, are not funnel-shaped or flaring at their mouths as are the estuaries that indent the so-called *ria* coasts of northwestern Spain and similar regions. Not only are the fiords relatively narrow throughout their length but they are fairly straight also. Fiords may also have many branches, extending inland like long canals. In fact, some of the fiords of Alaska are known as "canals," as for instance Lynn Canal at whose head is the town of Skagway. Fiord coasts likewise have many offshore islands, as if the entire edge of the mainland had been cut off and broken into pieces. Between the islands, likewise, are many narrow straits which are actually fiords also.

Some of the fiords of southern Norway are shown on the above map. The largest one shown, Sognefiord, penetrates more than 100 miles into the high, mountainous interior of Norway. Huge tourist liners ply far into the upper reaches of these fiords amidst most magnificent scenery, providing views of glaciers, distant snow fields, and rugged mountainsides. Waterfalls cascade down the cliffs from all directions. Small villages hide themselves in the heads of these long estuaries, almost isolated from the world. Little do we wonder, when we glance at a map like that above, that Norway has almost no railroads. Everywhere travel is largely by boat. From these lands came the Vikings first to Iceland, then to Greenland, looking for similar fiords, and then later to the shores of Cape Cod and even to New York and to Chesapeake Bay, some 500 years before Columbus.

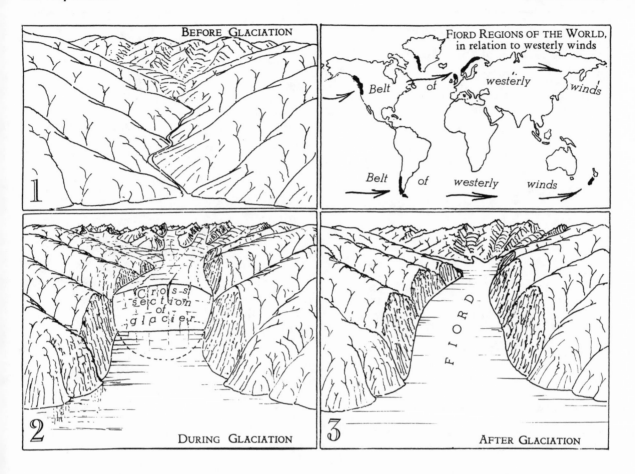

Fiords, as has been well known by geologists for some time, have been caused by glaciers, that is, rivers of ice which, coming down from mountainous areas, reach the sea. The fiords of the world are now largely free of ice, though in some cases small glaciers still exist in their upper portions, or even occasionally reach into tide-water.

A glacier, like a large river, flows down its valley. It is, however, immensely greater in depth than any river. Its channel therefore may be scoured far beneath the surface of the country, even far below sea level. It is this channel which, after the ice has melted away, constitutes the fiords.

In the illustration above three stages in the development of a fiord are indicated. In Figure 1 the region is shown before glaciation, with its river valleys. In Figure 2 the region is shown during glaciation, with the glacier plowing its way down the valley, and deepening as well as widening it. The glacier, in fact, gouges its channel well below sea level, so that in Stage 3, which shows the region after glaciation, the glacial channel becomes a fiord.

Because at their ends, where they melt, glaciers do not erode, fiords are shallower near their mouths than in their upper portions.

Note on the small map of the world above that all the fiord coasts occur in the two belts of prevailing westerly winds, and in every case are on the westward side of the land areas. In these locations there were in Glacial Time, as there still are, the heaviest precipitation and runoff. This runoff was, during the Glacial Period, mainly in the form of snow and ice. Permanent ice fields and glaciers still exist in all of these regions. These fiord regions of the world at the present day, with their milder climates, are still regions of unusually heavy rainfall, and at times of heavy snowfall as well.

EXAMPLE 23

The Problem

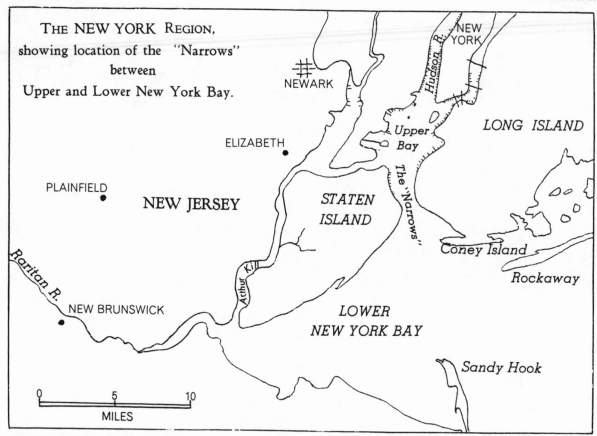

THE NEW YORK REGION,
showing location of the "Narrows"
between
Upper and Lower New York Bay.

NEWARK

ELIZABETH

PLAINFIELD

NEW JERSEY

Raritan R.

NEW BRUNSWICK

Arthur Kill

STATEN ISLAND

Upper Bay

The "Narrows"

Hudson R.

NEW YORK

LONG ISLAND

Coney Island

Rockaway

LOWER NEW YORK BAY

Sandy Hook

0   5   10
MILES

COAST LINES. *Straits. The Narrows, New York Harbor.*

The New York metropolitan area with its 550 miles of waterfront wharves is host to vessels of all nations. Coming through the 5-mile-wide opening between Sandy Hook and Rockaway Point, these vessels first enter the Lower Bay. Five miles offshore they have already got in touch with the Ambrose Lightship, which marks the seaward extremity of the Ambrose Channel. This is the point toward which all the steamship routes converge. Here they make their first stop, perhaps after several thousand miles of sailing, to take on their harbor pilot. They are still 20 miles from the southern tip of New York City.

Cruising slowly for the next 13 miles brings them to the Narrows, the entrance to the Upper Bay. Less than a mile wide, this narrow channel separates the western tip of Long Island from Staten Island. Fort Hamilton and Fort Wadsworth, respectively, guard the two shores. They are now of historical rather than of military significance. On both shores the land rises in low hills which extend inland, in Long Island to Forest Hills and beyond, and on Staten Island to Dongan Hills and Tottenville.

Beyond the Narrows the Upper Bay opens up, an expanse of protected water more than 3 miles wide and 5 miles in length. Of all the wonderful sights that the world traveler may enjoy, none is so impressive as the towering skyline of New York City as seen from Upper New York Bay. And to the emigrant or refugee from other lands it is truly an inspiring experience. At his left, coming up the Bay, he has the Statue of Liberty close at hand. Across the Bay, along the Brooklyn shore, are some of the many miles of wharves that line all the waterways around New York. And on the New Jersey side are the large piers where the railroads of the country can discharge their freight alongside the ships themselves.

The importance of the Narrows to the Port of New York is evidenced by the protected character of the Upper Bay. Here many ships of all kinds ride at anchor waiting for cargoes of one kind or another, or to dock somewhere around the city. The Lower Bay is far less protected and is much more remote from the centers of activity. Thus, the Narrows help to render the Port of New York one of the most remarkable harbors of the world. We can well ask what fortunate circumstance caused this constriction in New York Bay.

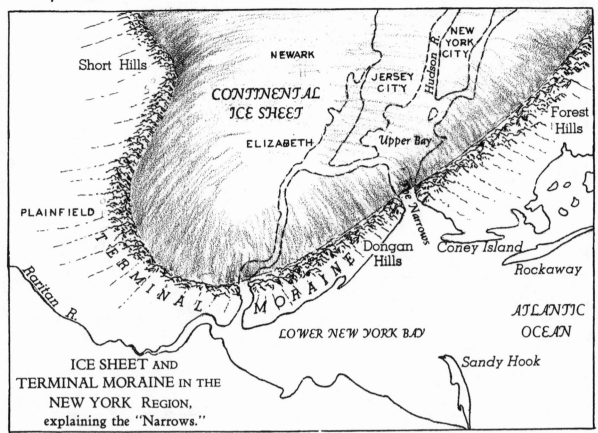

ICE SHEET AND
TERMINAL MORAINE IN THE
NEW YORK REGION,
explaining the "Narrows."

The Narrows, the narrow strait in New York Harbor, has few counterparts elsewhere in the world. Here the terminal moraine forms a belt of hills encircling the southern rim of the metropolitan area. This belt is breached by the Hudson River. A look at the above map shows that this moraine has been deposited across the center of New York Bay, thus cutting it into two parts, the Upper Bay and the Lower Bay. Only because of the great volume of the Hudson River has a channel across it been kept open at this point. Not only is this due to the strength of the currents, but here the glacial deposits which make up the moraine are easily eroded, much more so than solid rock.

If we follow this moraine much farther eastward, we find that it has been breached in several other places. There are of course the large interruptions in the moraine between Long Island and Block Island, and also between Block Island and Martha's Vineyard. But let us examine that portion of the moraine which forms the Elizabeth Islands off the Massachusetts coast. The morainal chain of the Elizabeth Islands is cut across by several passageways known as "holes." The best known of these is Woods Hole at Woods Hole, Massachusetts. This passageway is used by steamers plying between New Bedford and Martha's Vineyard, and is similar to the Narrows at New York. It is actually analogous to it in origin. Were we to look elsewhere in the world for such a feature along a coast, we should have to go to Denmark. There the long channel of Lim Fiord, actually a strait, cuts across the moraine at Aalborg, and separates the tip end of Denmark from the rest of the peninsula.

On the continents themselves moraines are commonly transected by river valleys. Were these valleys to become submerged by the sea, they would become straits. An example of this is Arthur Kill between Staten Island and the mainland of New Jersey. This is labeled on the opposite map.

EXAMPLE 24

The Problem

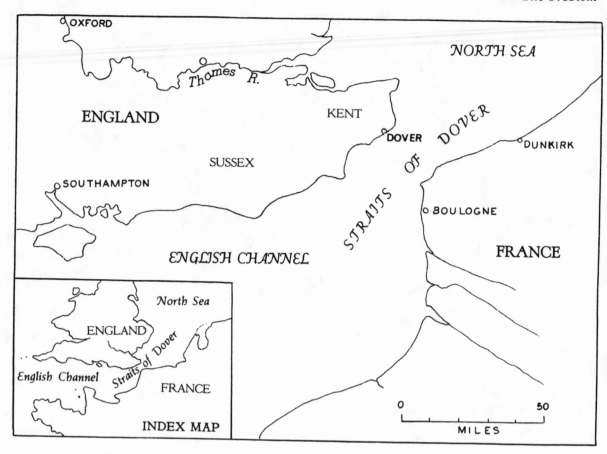

COAST LINES. *Straits. The Straits of Dover.*

Straits are narrow waterways between areas of land. They may also be thought of as constricted passageways connecting two larger bodies of water. We call to mind the Bering Straits, which connect Bering Sea with the Arctic Ocean and which separate North America from Asia; the Strait of Gibraltar, which separates Europe from Africa and connects the Mediterranean with the Atlantic; and the Strait of Bab el Mandeb separating Asia from Africa and connecting the Red Sea with the Indian Ocean. We think of the Malacca Straits, which separate Malaya from Sumatra and which gave name to the Straits Settlements at Singapore.

Man became acquainted with the straits of the world as a result of his early explorations, and he numbered them among his most important discoveries. No straits have been more important than the Straits of Dover separating England from France. England's relative isolation has depended upon this narrow body of water, only 20 miles or so across, and narrow enough to be crossed by even primitive boats. Across these straits time after time people from the continent have entered England. It was the focal point for the great Roman roads which converged here from both sides of the Channel.

It may seem strange that we think it worth while to wonder how the Straits of Dover came about. Our first explanation would be that England and the Continent were once connected and that this is where they broke apart. While this is strictly true, it nevertheless is hardly an explanation.

It is clear when we examine the straits of the world that there is no one explanation which takes care of them all. The Straits of Gibraltar and of Messina are similar in that they are both breaks across the South Europe Alpine Mountain System. Torres Strait between Australia and Tasmania is a down-dropped fault block. Perhaps a good approach to a solution for the Straits of Dover would be to discover why France and England are so close to each other at this particular point, and not worry about how they became separated.

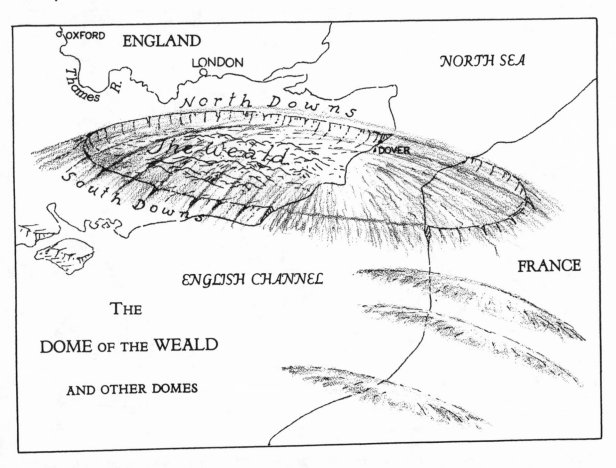

The above diagram is an effort to show how France and England were once joined with each other at the Straits of Dover. An elliptical dome-shaped arch of the earth's crust extended from southern England to France in a northwest-southeast direction. There were other elongated but smaller domes in France which did not reach across the Channel. From the flanks of the larger dome streams flowed toward the southwest to the English Channel, and toward the northeast to the North Sea. In other words this larger dome was the divide between two drainage systems.

At this point it would be proper for anyone who wishes to, to object to this fanciful explanation and to ask what evidence there is that a dome ever existed here. The answer of course is in the fact that the present-day remnants of the dome may actually be seen. The upper part of the dome has, at the present time, all been eroded away, but the roots of the dome remain. They form circular, or rather semicircular, features or ridges in the landscape of both England and France. In England, in the counties of Kent and Sussex, the remnant of the dome is known as the "Weald." In France a similar area is called the "Boulonnais," after the city of Boulogne.

The diagram above shows the semicircular ridges surrounding the Weald. On the north side of the Weald, just south of London, is a ridge of limestone, or rather chalk, some 400 feet high. Its northern slope toward London is gentle, conforming with the original slope of the dome surface. Its southern face is steep, being eroded across the rock structure. This northern ridge is called the "North Downs" in contrast with its continuation on the south side of the Weald called the "South Downs." The North Downs come to the coast in the White Cliffs of Dover, the South Downs at Beachy Head. The central part of the Weald is a hilly wooded area, the term "Weald" being a modification of the German word "Wald," meaning a woods. The Straits of Dover now cover the central part of the eroded Wealden dome.

EXAMPLE 25 *The Problem*

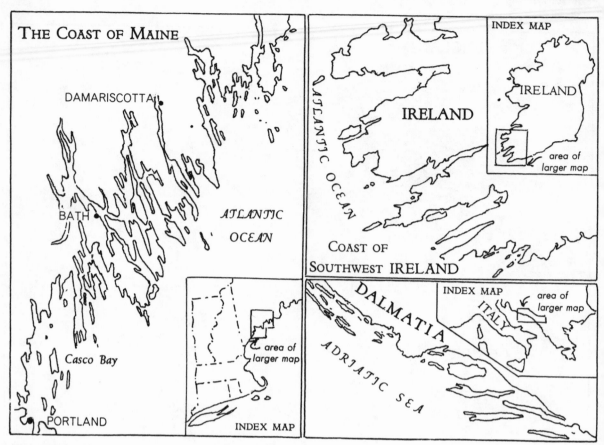

ISLANDS. *Linear Islands and Peninsulas ("Almost Islands"). The Maine Coast; Southwest Ireland; the Dalmatian Coast.*

The last few examples of shore lines have been leading gradually to a discussion of islands. Islands and peninsulas are of course closely related to each other. It takes very little to convert a peninsula into an island, and perhaps even less to convert an island into a peninsula.

A glance at almost any map will show that islands have many shapes. Some of these shapes, such as the linear examples illustrated above, may have real significance, whereas other shapes are merely fortuitous. Linear islands may be transverse to the general trend of the coast, as along the coasts of Maine and of Ireland, or they may be parallel to the coast, as they are along the eastern shores of the Adriatic. The islands which project outward from the Maine coast are usually continuations of peninsulas. In fact, anyone navigating such a coast without a chart would have difficulty knowing whether he was skirting an island or a peninsula.

Although the Irish coast does not display many islands, those which do occur are linear in form. It is easy to see that the Irish coast and the Maine coast belong in the same category. But our third example, the Adriatic coast, seems to be in a different class. The linear islands, as well as the occasional linear peninsulas, lie parallel to the coast line. Moreover, this coast line is hardly at all embayed. Nevertheless, the islands themselves appear to be very much like those in the other two cases.

In all of the regions illustrated above, the islands are closely related to their mainland areas, being detached only slightly. This is quite in contrast with many islands of the world that are oceanic in character. It can fairly well be inferred, therefore, that the geological conditions which occur on the mainlands of Maine, Ireland, and Dalmatia extend also into the offshore islands as well. It remains next to consider what these geological conditions are.

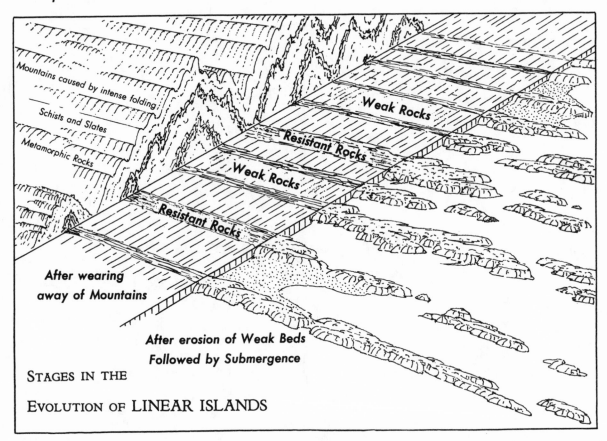

Mountains caused by intense folding

Schists and Slates

Metamorphic Rocks

Weak Rocks

Resistant Rocks

Weak Rocks

Resistant Rocks

After wearing away of Mountains

After erosion of Weak Beds Followed by Submergence

STAGES IN THE

EVOLUTION OF LINEAR ISLANDS

An attempt is made in the above diagram to show how an area like the Maine coast has been brought about.

In the back or top part of the drawing is represented a facsimile of the mountain structure that once prevailed in Maine. The folding as shown in the vertical cross section was intense and complex. There were numerous folds of different sizes, with much crumpling of the formations, so that individual beds in some cases became almost indistinguishable.

In front of the mountain area is shown a second zone or belt from which the mountains have been worn away, down to their roots, as it were. Here can be seen the various layers of rock forming parallel belts. Some of these belts consist of weak rocks; others are more resistant to erosion. However, all have been worn down in similar fashion during the long ages while erosion was taking place.

In the third and nearest zone the region is shown in its present condition. The weaker belts have been eroded out to form valleys, and the more resistant beds remain to form relatively flat-topped ridges. Many of the valleys have become submerged, as a result of changing sea level, and this has caused some of the ridges to become islands. The linear form of these islands is thus explained.

Whether the islands are transverse or parallel to the general outline of the coast depends entirely upon the direction of the rock structure. On the Dalmatian coast the islands, the peninsulas, as well as the ridges on the mainland, all run parallel with the coast line. But on the Maine coast and the Irish coast the transverse direction is more common. But even on these coasts there are places where the coast line swings around to conform with the trend of the rock structure.

A further discussion of the islands of the Dalmatian coast is given in Example 31.

EXAMPLE 26

The Problem

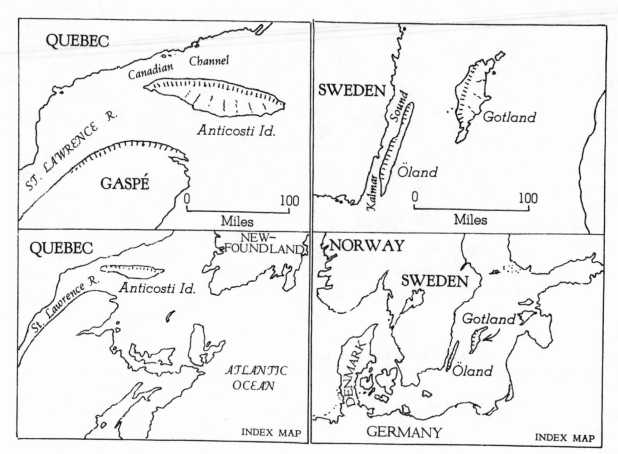

ISLANDS. *Linear Islands. Anticosti Island in the Gulf of St. Lawrence; Öland and Gotland in the Baltic Sea.*

These three large elongated islands, each approximately 100 miles long, are noteworthy because they occur almost singly and are not members of a "swarm," as in the preceding example. Moreover, most atlas maps provide a faint clue to their origin by indicating to some extent the character of their topography. As portrayed on the accompanying maps by means of hachures, each of these three islands consists of a steep scarp and a more gentle back slope. The scarp in each case, too, faces toward the adjacent land area, from which it is separated by a narrow strait or passageway.

On Anticosti Island the scarp has an elevation of several hundred feet, rising abruptly above the Canadian Channel at its base. Parts of the scarp crest attain elevations of 700 feet or so. The south half of the island is a broad sloping, almost unbroken, plain.

Öland Island, just off the coast of Sweden, is separated from the mainland by Kalmar Sound, which in places is only four miles across. Like Anticosti, it has a low scarp along its western edge, and otherwise is fairly level. It is more than ten times as long as it is wide, and is therefore much more linear in shape than is Anticosti.

Gotland is perhaps not exactly a linear island, although it is somewhat elongated. Because of its closeness to Öland and its similarity of form to Anticosti, and because of the fact also that it too is characterized by a scarp facing toward Sweden, it certainly belongs in the same category as the other two islands. Of the few other islands of the world which belong in this category Long Island, near New York, is an outstanding example. But this unique island has experienced certain profound modifications which almost put it in a class by itself. For this reason it is treated independently in the next Example.

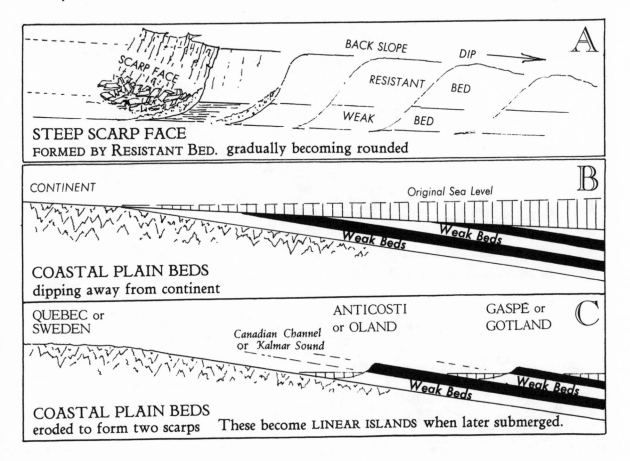

STEEP SCARP FACE
FORMED BY RESISTANT BED. gradually becoming rounded

COASTAL PLAIN BEDS
dipping away from continent

COASTAL PLAIN BEDS
eroded to form two scarps   These become LINEAR ISLANDS when later submerged.

It is a well known fact in geology that many asymmetrical ridges or scarps are underlain by resistant formations which dip in the direction of the back slope of the scarp. This is demonstrated in Figure A above. As the geologist expresses it, "The scarp face is developed across the structure and the upland above the scarp slopes gently in the direction of the dip of the beds." Some scarp faces are very steep indeed, being veritable cliffs like the Palisades of the Hudson. Others are merely gently rolling hills like the scarps that form the Chiltern Hills and the Cotswold Hills and other so-called "Downs" of England. Steep scarps are formed on extremely resistant and brittle formations like the "traprock" of the Palisades. The Downs of England, however, are mostly chalk or weak limestones. The almost vertical joints or cracks in the resistant rocks cause the blocks produced by weathering to break off along vertical planes.

Coming now to the two regions discussed on the preceding page, we note that in each case there are two more or less parallel scarps: Anticosti and Gaspé in the one case, and Öland and Gotland in the other. This means that two resistant formations are represented in each of the two areas.

The past history of each of these two regions is shown simply in Figures B and C above. In B is represented a series of formations dipping gently toward the right, and resting upon a basement of more complex rocks which outcrop on the mainland to the left. The series of dipping beds, originally laid down under the sea, constitutes a "coastal plain" when the sea is withdrawn. The erosion of this coastal plain gradually produces the features shown in Figure C. The weaker beds have been eroded away to leave the relatively resistant beds to form scarps. Slight submergence has then caused the scarps to become islands, linear islands like Anticosti, Öland, and Gotland.

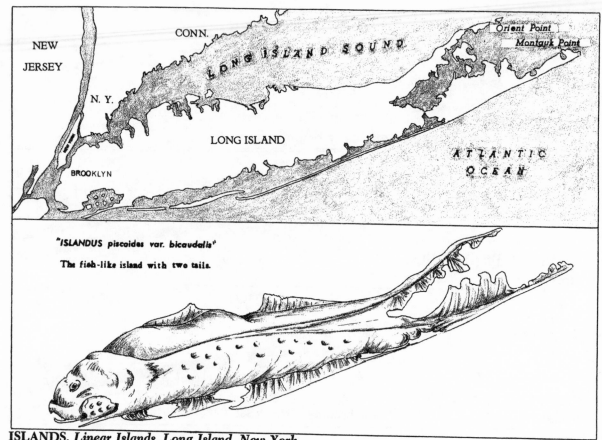

"ISLANDUS piscoides var. bicaudalis"

The fish-like island with two tails.

ISLANDS. *Linear Islands. Long Island, New York.*

Like a great fish, Long Island lies off the Atlantic Coast, headed toward New York City as if to prey upon its teeming life. The piscoidal form of Long Island is shown in the sketch below the map. In this sketch I have allowed my imagination to have a little play, with somewhat fearful results. Even more fearful is the scientific name which I felt it was my privilege to bestow upon it.

Long Island is another example of a long-shaped island, and its name is therefore very descriptive and appropriate. Long Island fundamentally is very much like the islands of Öland and Gotland, at least in origin. An explanation for the origin of Long Island is given later in Example 37, in connection with the development of the drainage features of the Atlantic Seaboard. It is, in short, what is called a cuesta, its foundation being part of the bedrock of the Coastal Plain. At this point, however, our attention is to be directed to those two tail-like appendages which stretch out from its eastern end. The northern tail forms Orient Point, to the east of which is the chain of the Fishers Islands along the Connecticut coast. The southern tail forms Montauk Point, and east of Montauk and just within sight of it on a clear day is Block Island. Between Orient Point and Montauk Point is the embayment known as Great Peconic Bay. The largest island in this bay is Shelter Island. As for the other irregularities around the coast of Long Island, these will not engage us at this moment, for as we shall soon see they are not related in any way in origin to the two eastern tails.

As we puzzle over these features and think of other irregular coasts which we have seen, it becomes fairly clear that the indentations making up Peconic Bay are not drowned river valleys such as those along Chesapeake Bay and the coast of Spain described in Example 21.

At the western end of Long Island between Brooklyn and Staten Island, at the corner of the map, is the narrow entrance to New York Harbor known as the Narrows. Through these narrows pass all the great liners which enter the Port of New York. Our explanation, we shall find, will explain this feature as well as the two tails at the eastern end.

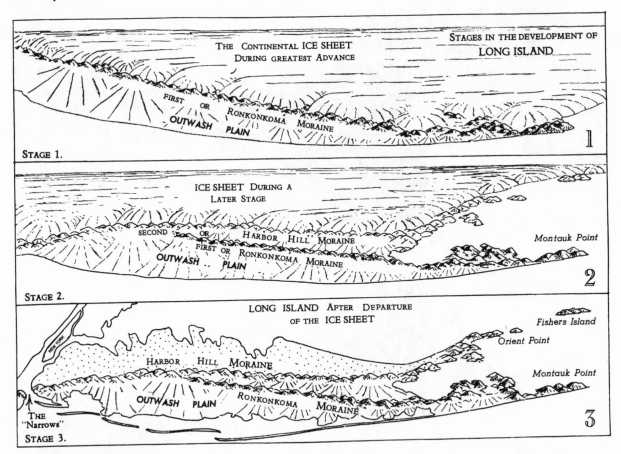

The features of Long Island discussed on the preceding page may readily be understood by consulting the three diagrams above. Going back to Glacial Time, we see first (Figure 1) the great ice sheet at the time of its most southerly advance. For a fairly long period of time the front of the ice remained more or less stationary. As a matter of fact what really happened was that it melted away just as fast as it moved forward. There was thus a constant transfer of the load of debris carried by the ice to the ice front. Here it was piled up to form an irregular line of hills known as a "terminal moraine." To this particular moraine the name "Ronkonkoma Moraine" has been given. Lake Ronkonkoma, a large lake on Long Island, lies in a "kettle hole" in the middle of this moraine.

The next stage (Figure 2) represents the position of the ice front after the ice had melted back and then moved forward again to a position somewhat north of the previous one. Again a moraine was formed. This moraine is called the "Harbor Hill Moraine."

Now it can easily be seen (Figure 3) that it is these two moraines which form the two tails of Long Island. The Ronkonkoma Moraine forms Montauk Point; Harbor Hill Moraine forms Orient Point, together with various islands in Peconic Bay.

Coming now to the western end of Long Island, we see in Figure 2 that the second advance of the ice actually extended farther south than the first advance. This obliterated the first moraine in western Long Island. It is this second moraine which runs through Brooklyn, where it may be seen in Prospect Park, and which extends over into Staten Island. The narrow passageway through this moraine which almost blocks up New York Harbor forms the entrance to Upper New York Bay. This feature has already been discussed in Example 23.

EXAMPLE 28   *The Problem*

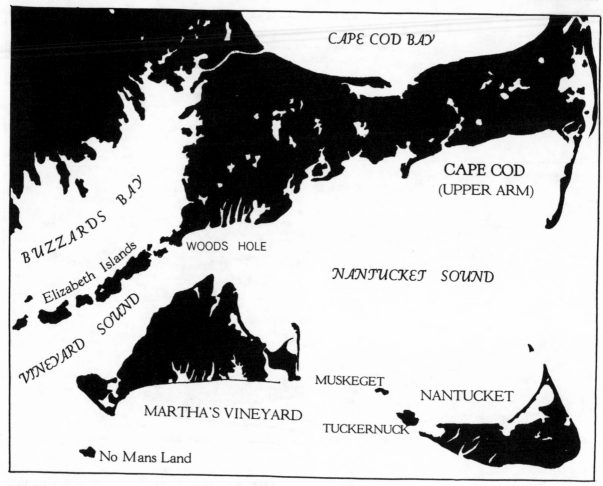

ISLANDS. *Arcuate Islands. Martha's Vineyard and Nantucket.*

Among the unusual islands of the world, Martha's Vineyard and Nantucket are almost in a class by themselves. There are very few, if any, islands quite like them. Let us therefore note some of their characteristics.

Martha's Vineyard is shaped something like a triangle, with its base toward the south and its apex pointing north toward the mainland. Its eastern and western sides are somewhat curved, the eastern perhaps more noticeably than the west. Nantucket is also strongly arcuate, being almost a counterpart of the eastern half of Martha's Vineyard. Off the southwest corner of Martha's Vineyard is the little island of No Mans Land, just as off the western end of Nantucket are the islands of Tuckernuck and Muskeget. The southern shore both of Martha's Vineyard and of Nantucket is a long simple beach which cuts across the ends of a series of lakes that look like drowned river systems.

Among other things which may fit into this picture are the features on Cape Cod, that is, the so-called upper arm of the Cape. Cape Cod is strongly curved, and is characterized by many small lakes and ponds. It has also on its southern shore several long parallel embayments which resemble the drowned rivers of Martha's Vineyard.

From the western end of Cape Cod, at Woods Hole, there extends out westward for 20 miles or so a chain of islands—the Elizabeth Islands. These separate Buzzards Bay from Vineyard Sound. They also have a curved form.

This is only a fragment of the whole picture, for much more lies to the west. There is enough, however, to provide an interesting story.

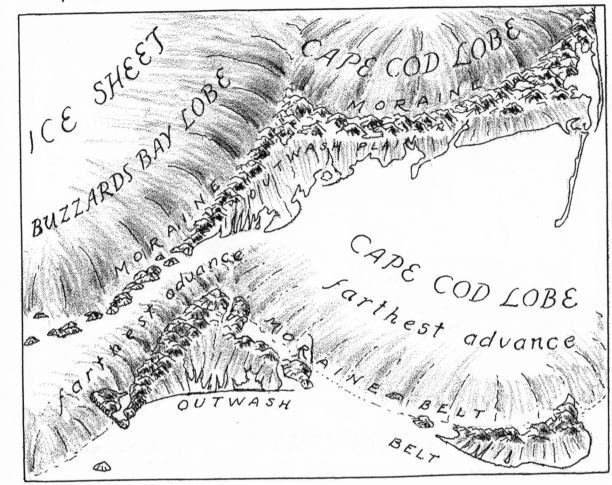

Martha's Vineyard and Nantucket were formed by two lobes of the continental ice sheet. The front of one lobe (called the Buzzards Bay lobe) ran along the western side of Martha's Vineyard. The front of the other lobe (called the Cape Cod lobe) ran along the eastern side of Martha's Vineyard as well as the northern side of Nantucket. The hilly topography of both of these islands represents the terminal moraine. The outlying islands of No Mans Land, Tuckernuck, and Muskeget are also parts of the moraine. They are, however, pretty well reduced in size because of wave action.

Following the period during which these moraines were laid down, the ice retreated by melting back, and remained for a time in a more northerly position. The moraine deposited during this later stage is represented by the chain of the Elizabeth Islands, by the peninsula at Woods Hole, and by the long curving upper arm of Cape Cod. On Cape Cod many lakes lie among the morainal hills.

In front of each of the moraines are low sloping sandy outwash plains. The channels of the streams which formerly flowed across these plains from the ice front are in many places now submerged or drowned to form long lakes or lagoons. These are usually called "ponds" by the local people. Many other channels not now containing water crease the surface of these plains.

Since the deposition of the moraines and outwash plains the waves and currents have built many spits and bars. The most notable of these is Monomoy Point, which extends south for many miles from Cape Cod.

The moraines of Martha's Vineyard have been laid down upon a foundation of older beds, mostly clays. These may now be seen in the brightly colored cliffs of Gay Head at the western end of the island.

EXAMPLE 29

The Problem

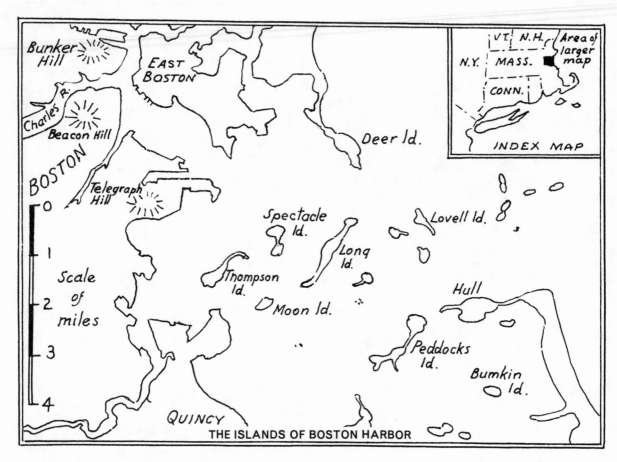

THE ISLANDS OF BOSTON HARBOR

ISLANDS. *Elliptical Islands. The Islands of Boston Bay.*

In Boston Harbor is an unusual display of small islands. A notable feature about all of these islands is the tendency to have an elliptical shape. Some of the larger islands appear to consist of two ellipses joined together. Spectacle Island derives its name from a shape of that kind. Other islands, like Peddocks Island, seem to be formed of several ellipses joined to each other. So also does Long Island.

Not only are the islands elliptical but parts of the mainland likewise tend to be of that form. Some of the peninsulas projecting from the mainland are merely islands attached to the land by a sand bar. This is true of Deer Island and of the peninsula of Hull.

Most of the islands have small projections jutting out in one direction or another. These are doubtless sand spits of some kind.

There are several hills on the mainland, too, which have elliptical forms, and which are more or less the same size as the islands. Bunker Hill and Beacon Hill, upon which the State House stands, are well known examples. Telegraph Hill is another example. There are several others less well known and not shown on the above map.

No other large harbor has islands of the kind which Boston has. The islands in Casco Bay at Portland, Maine, for example, are very different. Most of them are long and irregular.

The Esso automobile road map of southern New England includes an enlarged section showing Metropolitan Boston. The above map is reproduced on the same scale as the road map. It will be noted that most of the islands are much less than a mile long. Some in fact are mere rocks, occasionally surmounted by a lighthouse. The Index Map indicates the location of the Boston area with relation to the rest of the New England coast. No other part of the New England coast resembles the Boston region.

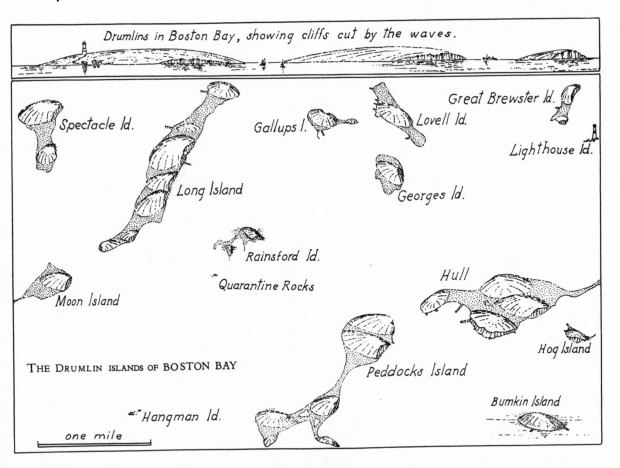

Drumlins in Boston Bay, showing cliffs cut by the waves.

THE DRUMLIN ISLANDS OF BOSTON BAY

The above sketch map is modeled after part of the United States Geological Survey topographic map of Boston Harbor. It has been reduced to about half its original size. An attempt has been made in this reproduction to represent the topography as it would look from an airplane. It shows that all of the islands are made up of elliptical-shaped hills. Connecting the hills in many places are bars of sand. The sand areas are stippled on the above map. All of the islands are utilized in one way or another, and many of them contain government buildings of some kind. A few of the piers projecting into the water are indicated, but none of the buildings.

Some of the islands have fantastic shapes, looking for all the world like animals. Bumkin Island, for example, resembles a turtle.

Many of the islands have been cut off by the waves so that the ellipses are not complete. It is interesting to note that most of the cliffs face to the east, toward the open ocean. It is from this direction that the strongest waves come. Several of the smaller islands have been completely destroyed by the waves, and are represented now only by piles of rocks. Such are Quarantine Rocks and Hangman Island.

The peculiar elliptical hills which make up the islands in Boston Bay are known as "drumlins." They were formed by the continental ice sheet. They consist of loose soil and rocks, and in that respect are very much like a moraine. Unlike a moraine, however, they are smooth and rounded because the ice once passed over them.

Being of loose material, the drumlins are easily eroded by the powerful waves which strike this part of the coast. The small sketch above the map gives the appearance of some of these islands as seen from water level.

In New York State, south of Lake Ontario, there is another notable drumlin "swarm." Many of the drumlins have been cliffed by the lake, just as they have been in Boston Bay.

EXAMPLE 30

The Problem

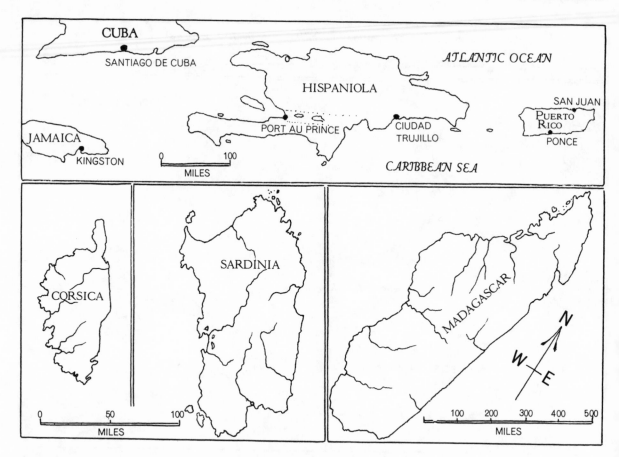

ISLANDS. *Rectangular and Straight-Sided Islands. Puerto Rico, Cuba, Haiti; Corsica and Sardinia; Madagascar.*

Many coast lines of the world exhibit straight-line segments. This is particularly true of islands, notably the fairly large ones. Some, in fact, like Puerto Rico, are almost rectangular in form. Puerto Rico, the easternmost of the Greater Antilles, is only one member of this east-west chain of larger islands, all of which display here and there straight-sided coast lines. These linear elements, it is worth noting, all trend in general in an east-west direction. The several promontories on the island of Hispaniola (often called Haiti) emphasize this in a striking manner. And this trend of features conforms also with some of the details in the interior of the island itself. Not only this, but the straight side of the northernmost promontory of Hispaniola is continued westward along the southern shore of Cuba. Jamaica as a whole extends in an east-west direction in conformance with the other Antilles. The details of its shore line, however, are relatively short, and are not emphatically straight-sided.

Coming next to the relatively large islands of Corsica and Sardinia in the Mediterranean, we can detect certain resemblances with the islands of the West Indies. Sardinia is fairly rectangular, and is reminiscent of Puerto Rico. It is about the same size too, perhaps a trifle larger. Corsica has one long straight-sided peninsula which resembles those on Hispaniola.

Of all the islands of the world, Madagascar exhibits the longest unbroken stretch of straight coast line, well over 500 miles in length. The western side of Madagascar, however, is far from straight. In fact, its northern portion is minutely embayed. Corsica, like Madagascar, also has a very irregular western coast with many small embayments. The problem before us, therefore, is to explain not only the straight-sided coasts but also the many departures from this condition on the same islands.

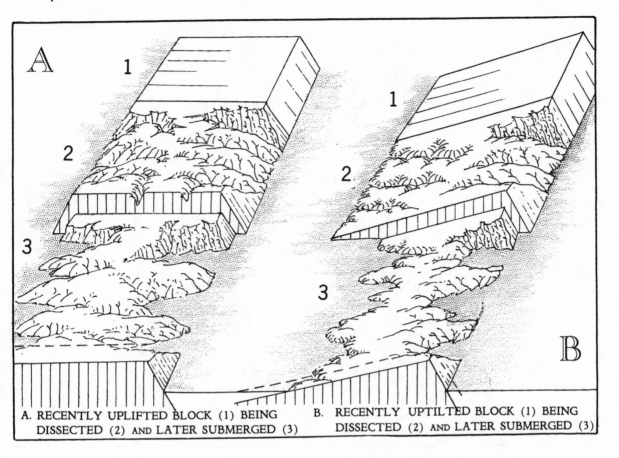

A. RECENTLY UPLIFTED BLOCK (1) BEING DISSECTED (2) AND LATER SUBMERGED (3)　　　B. RECENTLY UPTILTED BLOCK (1) BEING DISSECTED (2) AND LATER SUBMERGED (3)

This explanation is designed to show why some islands are straight-sided, and also why parts of these same islands are not.

Let us take in the first place an uplifted block of the earth's crust bounded on all sides by fault lines or fractures. Such a block is shown in Figure A, 1, above. In its initial stage this block is apt to be roughly rectangular. Its sides are fairly clean-cut and simple in outline. However, in nature, this condition will not prevail for long. Stream dissection sets in, and the island becomes hilly or even mountainous in character. Its sides, nevertheless, remain fairly straight, as in Figure A, 2. This presumes a condition of stability. Such a state of affairs is hardly to be expected, however, in an island that has already shown a tendency to rise above the general level of the earth's surface. The zones of weakness, that is to say the fault lines by which it is surrounded, permit it to rise again or to subside again with changing stresses in the earth's crust. A slight subsidence, as shown in the foreground of Figure A, results in the partial drowning or submergence of the valleys. The coast line thus becomes irregular or embayed. These later movements, whether of subsidence or of renewed uplift or of tilting, all change the initial outline of the island and obscure or even obliterate its original straight shore lines.

Tilted blocks, too, as shown in Figure B, frequently tend to be fairly straight on one side and quite irregular on the other. The reason for this is that when a change in elevation occurs the high side is not altered much in outline by a change in the elevation of the shore line. The low back slope, however, becomes irregular by only a slight subsidence of the land, and large irregularities result. This has happened on the western sides of Madagascar and also of Corsica.

For other straight-sided features it may be interesting to refer again to the description of Korea in Example 5.

EXAMPLE 31

The Problem

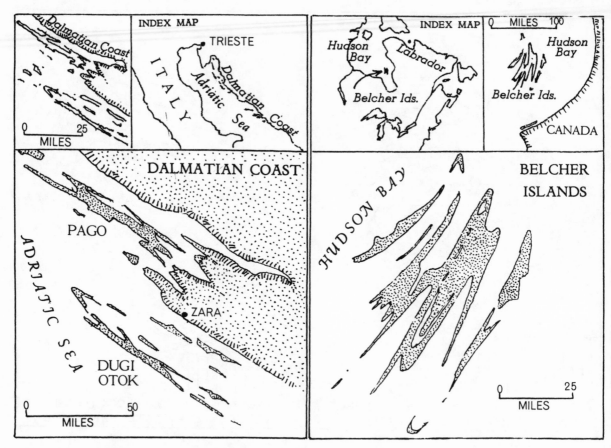

ISLANDS. *Zigzag Islands. The Dalmatian Islands, Adriatic Sea; Belcher Islands, Hudson Bay.*

Among the various island forms none are quite so apt to capture our attention as those which are forked in various ways, and those which have a zigzag pattern. Although there are many cases of single islands which are forked or pronged, there are only a few cases of extensive island groups which reveal such strikingly zigzag features as the two examples shown on the above maps.

At the left above are shown in several different scales the Dalmatian Islands of the eastern Adriatic. These form a chain which runs parallel to the coast of Yugoslavia for more than 250 miles from Fiume southward. Because the small scale of most maps reveals very inadequately the actual shape of the various islands of this chain, there is given also, in the sketch map above, a somewhat idealized presentation of the smaller details. The islands of Dugi Otok and Pago deserve particular attention, as well as adjacent parts of the mainland of which these islands are merely detached fragments. It is quite obvious that the forms of these islands are by no means fortuitous, but that they are controlled by some fundamental geological structure.

The other group of islands shown above, the Belcher Islands of Hudson Bay, is even more remarkable than the Dalmatian group. The largest island in the Belcher group, called Flaherty Island, displays a number of almost perfect zigzags. Not only that, but note also that its zigzag features are continued into the adjacent islands on either side. These several islands are doubtless all parts of one large geological structure.

Long before the Canadian Government produced its excellent topographic map of these islands, through the medium of aerial surveys, the Belcher Indians who inhabit them had prepared on skins remarkably accurate drawings to guide them in their canoe trips through the many intricate passageways.

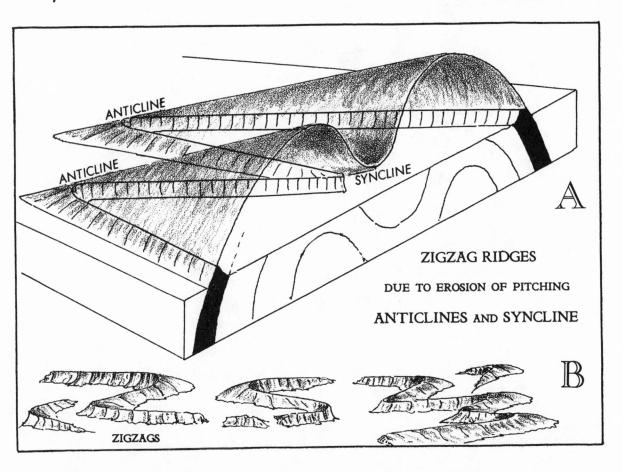

ZIGZAG RIDGES

DUE TO EROSION OF PITCHING

ANTICLINES AND SYNCLINE

Zigzag features occur not only on islands but on the mainland as well. In many parts of the world the hills are in the form of long ridges which zigzag back and forth across the country. The rivers occupying the valleys between them have a corresponding zigzag pattern. In places along the coast, where the valleys become submerged, the hills stand up as linear promontories or islands having hooks or angular bends.

The transition from zigzag mainland ridges to zigzag islands may be clearly observed along the Adriatic coast. The explanation for these features, whether on the mainland or on the islands, is found in the geological structure of the region.

Without attempting to explain the fairly complicated structure of either the Dalmatian or the Belcher Islands, let us instead take a very simple case of zigzag ridges such as that shown in Figure A above. Here we see two uparched folds, known as "anticlines," which are sloping down or "pitching" toward the left. The down-warped part between the anticlines is a "syncline." The original form of these folds before erosion is indicated by the shaded part of the drawing. After erosion only the root of the original fold remains, and this has a zigzag pattern. With a piece of paper and shears this can readily be demonstrated.

In Figure B above is shown a somewhat more complicated set of zigzags. These resemble a little the islands of the Adriatic and of Hudson Bay. It is quite possible to determine from the appearance of the sketches which parts of these islands are the roots of the anticlines and which parts are the roots of synclines.

Zigzag ridges are wonderfully displayed in central Pennsylvania. These are portrayed on the Explanation map for Example 43. It is easy to imagine what this area would look like, were it to become partially submerged.

EXAMPLE 32   *The Problem*

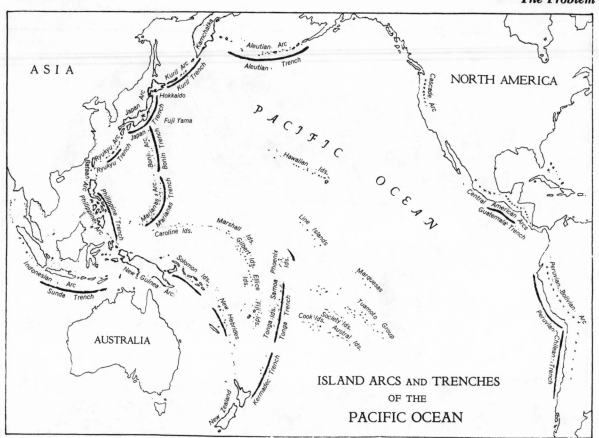

ISLAND ARCS AND TRENCHES
OF THE
PACIFIC OCEAN

ISLANDS. *Island Chains or Island Arcs. The Aleutians and Other Island Arcs of the Pacific.*

An examination of many maps shows that the islands of the world belong in two main categories: first, those which are essentially parts of the continents to which they are contiguous, such as Long Island, Newfoundland, and the British Isles; and second, those which form offshore or oceanic chains, extending in some instances from one continent to another. In addition, there are of course some isolated islands far removed from any other land.

The term "island arc" has been used to describe these great festoons of islands, a term that is especially appropriate to the islands of the western Pacific. Many of the arcs spring from continental peninsulas, and to that extent reveal their close relation to the continents; but they reach far offshore, 1,000 miles or more, and become real oceanic islands. By far most of the islands of the world belong in the category of the "island arcs."

On the above map a succession of island arcs can be followed all the way from Alaska to New Zealand. Some of the islands making up the arcs are big, almost continental in character. Others of the chain are mere pinpoints on the map. The Aleutian chain springs from the Alaskan Peninsula and swings in a southerly arc for much more than a thousand miles to the Kamchatka Peninsula, a projection of which juts out to meet it. From Kamchatka another chain, the Kuril Islands, runs south to the island of Hokkaido, Japan, to meet one of that triangular island's apices. Then southward the arc follows the Japanese archipelago, from which it continues to Formosa through the Ryukyu chain. Thence southward all the way to New Zealand, through the Philippines, New Guinea, the Solomons, and the New Hebrides, this great series of arcs continues. Throughout this belt practically all of the islands are surmounted by volcanic peaks, many still active.

A second roughly parallel arc springs from Japan exactly where Fuji is situated, running southward through the Bonin chain, the Marianas, the Caroline Islands, the Marshall, Gilbert, and Ellice Islands, to Fiji. From the northeastern promontory of New Zealand a series of loops springs northward to the Phoenix Islands, and still other loops extend eastward through Samoa, the Society Islands into Tahiti, and the Tuamotu archipelago.

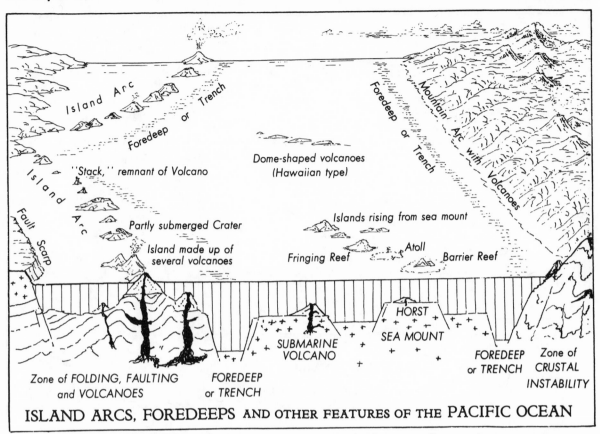

ISLAND ARCS, FOREDEEPS AND OTHER FEATURES OF THE **PACIFIC OCEAN**

There are two noteworthy aspects about the island arcs of the western Pacific which appear on the preceding map. First, it is to be noted that all of the arcs, both the island arcs and the "mountain arcs" of Central and South America, are convex, or bow out toward the Pacific Ocean. They appear to be anchored at certain nodes and to be bent out away from the continents where they stretch between these anchors.

In the second place, note that on the seaward side of each arc is an oceanic "deep," or trench. These two features of the island arcs, combined with the fact that they are all volcanic in character, have received abundant study by geologists. Certain mountain ranges of the world, like those of Central America and the Andes, are also surmounted by high volcanoes, and are very much like the island arcs. Other ranges, like the Alps and the Caucasus, as well as our own Appalachian System, are merely great areas of disturbance accompanied by folding and overthrusts, but with no volcanoes. Only in certain places along the margins of such disturbance are a few volcanic phenomena to be found.

From these facts it is clear that the island chains of the several oceans have had a past history not unlike that of some of the mountain ranges of the world. They represent zones of crustal deformation and of weakness. They are zones of fracturing and of faulting along which molten lavas from beneath the earth's surface have been able to rise.

Associated with the island chains are oceanic "deeps," or trenches, presumably down-dropped blocks, or grabens, of the earth's crust, as illustrated above. These trenches may have subsided because of the removal at depth of the great volume of material used in building the volcanic chain.

The islands themselves vary greatly in form. Some consist of single volcanic peaks, others of several volcanoes fused together. Still others are merely the eroded remnants of volcanoes, the result of wave action, mere "stacks" projecting above the sea like the so-called "Jimas" of the Bonin chain. There are still others which never reach the surface or which serve as the foundation for the many coral reefs of the South Seas.

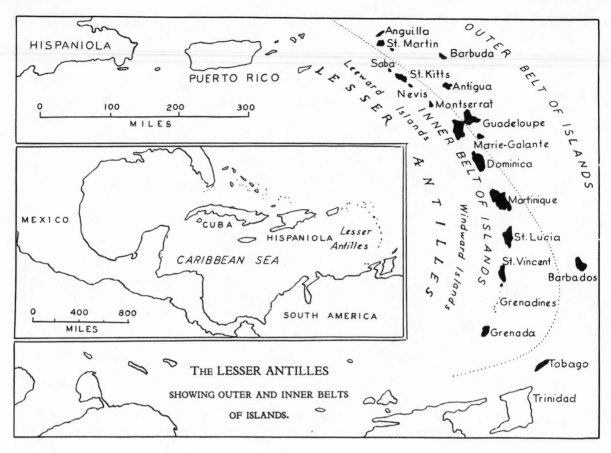

THE LESSER ANTILLES

SHOWING OUTER AND INNER BELTS

OF ISLANDS.

ISLANDS. *A Double Chain. The Lesser Antilles.*

On the average map of North America the chain of the Lesser Antilles looks like a series of minute stepping-stones which connects North and South America. Indeed, to some extent it serves this purpose to the migrating birds, although actually most of them, after coming down the Atlantic coast, go across to Yucatan.

These mere dots on the map, however, when examined on a larger scale, are found to fall into two belts parallel to each other. The inner belt contains the most numerous as well as the larger islands. Included in this belt are the well known French islands of Martinique and Guadeloupe, as well as the British islands of Dominica, St. Vincent, St. Lucia, and Grenada, not to mention several smaller ones. The outer belt, as is evident from the above map, includes St. Martin, Barbuda, Antigua, and Barbados. It includes also the eastern half of Guadeloupe and the adjacent small island of Marie-Galante. At its southern end this outer belt extends to Tobago and Trinidad. The mere fact of recognizing these two belts may perhaps be the first step in learning the names and the locations of these elusive spots on the map.

Another confusing matter may possibly also be clarified; namely, the use of the terms "Leeward" and "Windward" islands. It would seem in this breezy part of the world where the Northeast Trade Winds are always blowing that each island should be called "windward," or at least that each island should have a windward and a leeward side. This latter fact of course is true. We can perhaps understand the use of the term "Leeward Islands" for Guadeloupe and St. Kitts because they are on the leeward or inner side of the chain as compared with Antigua and Barbuda, which are on the windward side. The southern members of the arc of the Lesser Antilles are less protected by an outer belt, and therefore the term Windward seems well suited to them. At the present time the term Leeward Islands applies to all the islands in the northern half of the chain regardless of their windward or leeward position.

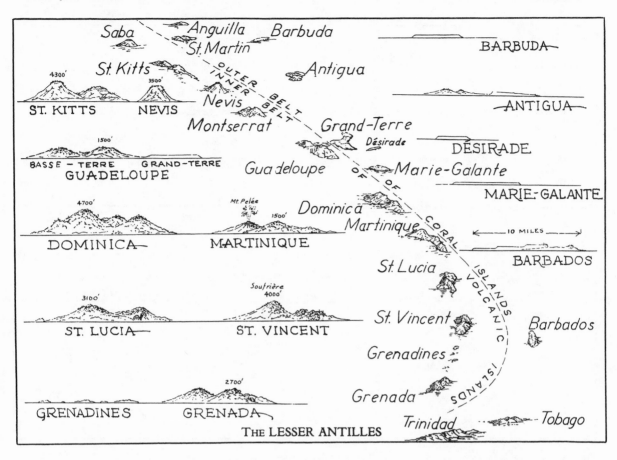

THE LESSER ANTILLES

The recognition of the two belts of the Lesser Antilles, that is, the "inner" belt and the "outer" belt, suggests also that the two belts probably differ from each other. That is true. The members of the inner belt are everywhere volcanic in form. Almost every island in this belt exhibits an active volcano or one that has recently been active. Some very beautiful cones may be seen on these islands. Mount Pelée on Martinique and Soufrière on St. Vincent are the ones that have been most recently active. There is also a Soufrière on St. Lucia. The minute island of Saba is a miniature cone with a crater in which there is now a little village of boatbuilders and fishermen. On Grenada is a volcano with a crater containing a crater lake, the Grand Étang, a point of interest to visitors. In a zone of this kind, earthquakes too are common.

The outer belt of islands is built on a foundation of old distorted and in general nonvolcanic rocks. On a number of these islands, such as Antigua and Barbados, these old rocks are visible, in some cases forming low mountains. Resting on these older rocks are coral reefs and limestone beds now raised above sea level. Nine-tenths of the area of Barbados is a limestone plateau rimmed by terraces which are wave-cut platforms representing successive stages in the uplift of that island. On Antigua about half the island consists of a dipping limestone plateau or "coastal plain" exhibiting a cuesta scarp and "inner lowland." The eastern half of Guadeloupe is a low and flat limestone plain barely attached to the high western volcanic area. Marie-Galante is a low limestone plateau believed to be an elevated coral atoll. Tobago and Trinidad, at the southern end of the outer belt, consist of old disturbed formations like those which underlie the island of Barbados and similar to those which are found on the mainland of South America.

From all this we conclude that the inner volcanic belt is a narrow line of weakness in the earth's crust along which volcanism has occurred. It is a fractured zone in the main belt of mountain disturbance which formed the Lesser Antilles as a whole.

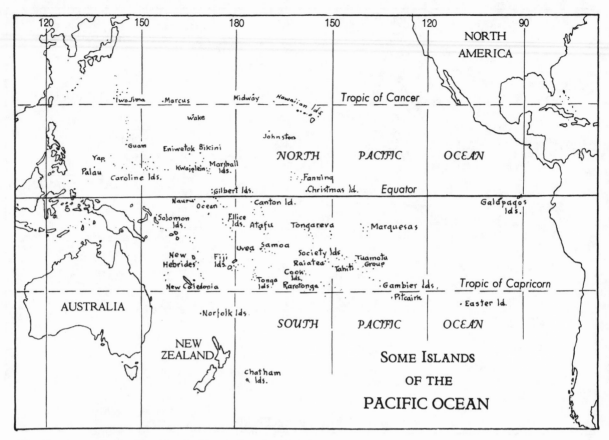

ISLANDS. *Oceanic Islands. Atolls of the Pacific Ocean.*

In the South Pacific Ocean, scattered over an area several times that of the United States, are literally thousands, probably tens of thousands, of small islands. They resemble on a map the stars of the Milky Way. Even on large-scale maps they are mere points. The traveler flying over the vast reaches of the Pacific for hour after hour sees only the limitless ocean. When he enters one of the great island swarms he hardly even realizes it, for he sees only one island at a time, so remote are these pinpoints of land from one another.

On maps of the Pacific Ocean, even fairly large-scale ones, these various islands, the large as well as the small ones, appear as mere specks. But if we take the trouble to examine some of the large-scale representations of these islands, as they are shown around the border of the National Geographic Society maps of the Pacific, and also in some atlases such as Bartholomew's, we note that these islands have many shapes. One thing they all seem to have in common is reefs of one sort or another. The splendid National Geographic Society map of the Pacific Ocean published in 1952 depicts in detail threescore of these various islands, presenting a full range of types. From these as well as from the thousands of other examples that appear on the Admiralty charts of the great maritime nations of the world it is possible to detect in these many island forms a certain systematic grouping of shapes based upon the character of the reefs which surround the islands.

It is possible roughly to classify these islands into three categories in the manner shown by the sketch maps on the following page. In the first category are those islands like Tahiti which are surrounded by a so-called "fringing reef," one which lies close to shore and which at low tide appears to be part of the island itself. The second type is represented by certain islands in the Society Group, such as Raiatea. In this type there is also a reef, but instead of hugging close to the island it lies well offshore to form what is called a "barrier

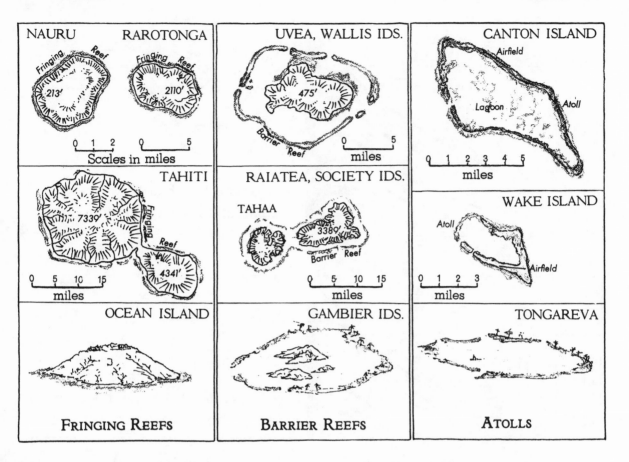

reef." Openings here and there in the barrier reef make it possible for ships to gain entrance into the broad circular lagoon of quiet water which encircles the island itself. The third category is represented by Canton and Wake Islands, both of which are steppingstones for planes flying across the Pacific. These islands are merely reefs or "atolls," circular bits of land surrounding a broad lagoon of usually shallow water. The map on the preceding page shows the location of these islands and some of the other island groups of the Pacific.

Among the other islands of the Pacific which belong to each of the three above-mentioned categories are the following: Like Tahiti with its close fringing reef are most of the islands of the Fiji group. Nauru and Ocean Islands, illustrated above, are similar examples, remote from any other islands. Nauru, a trust territory of Australia, is an important source of phosphate derived from bird guano. The island of New Caledonia is also surrounded by a fringing reef, which, however, in places lies well offshore and therefore appears to be a transition to the barrier type.

One of the most interesting of the barrier-type islands is represented by the Truk Island group. A score or more of islands, some a thousand feet or more in height, are surrounded far offshore by an almost continuous circular barrier reef which served during the war as a defensive wall to the islands lying within. Ponape Island in the Caroline group is another nice example.

The "atoll" type is represented by many examples, such as Tongareva, Fanning, Atafu, and the well known Eniwetok, Kwajalein, and Midway Islands of wartime renown.

The problem before us now is to explain this systematic arrangement of island forms.

EXAMPLE 34   *The Explanation*

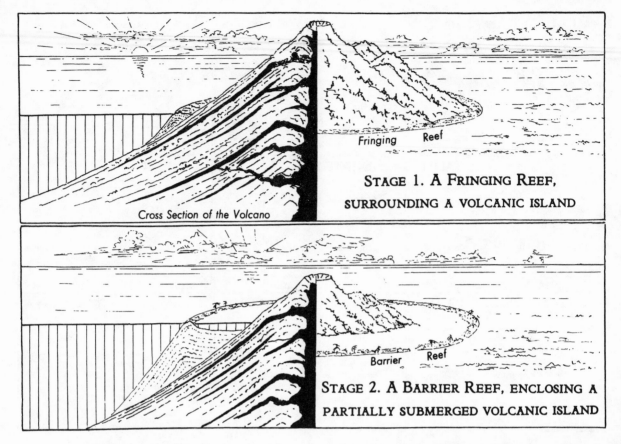

Cross Section of the Volcano

Fringing   Reef

STAGE 1. A FRINGING REEF,
SURROUNDING A VOLCANIC ISLAND

Barrier   Reef

STAGE 2. A BARRIER REEF, ENCLOSING A
PARTIALLY SUBMERGED VOLCANIC ISLAND

In the study of landscape when we encounter a sequence of transitional forms which merge gradually from one to another, we suspect that we are concerned with an evolutionary series. The forms are related to each other by some continuing process of development. This is unusually well exemplified by the several types of coral reefs which have just been presented. The successive stages in the evolution of these reefs from the fringing type through the barrier stage to the atoll are illustrated in the diagrams above.

As we have seen in the preceding examples, there are numerous volcanic islands scattered about the oceans of the world, not entirely hit and miss, but arranged more or less in belts. Doubtless an equal number of volcanic peaks rise above the ocean floor but do not protrude above sea level. A similar situation would exist if the volcanoes of the continents were submerged. Some would project, others would be covered by water. We know, too, that from time to time such volcanic islands come into being, born before our very eyes. We know that the continents, as well as the sea floor, are not perfectly stable, but here and there change their elevations, rising or falling through the ages. Indeed, these changes may be rapid enough to be observed and measured during historical time. It is not entirely surprising, then, to conclude that many of the isolated oceanic peaks are rising and that others are sinking. It is those which are sinking with which we have to deal in this example.

Stage 1, illustrated above, shows an island which has remained stable enough to permit the growth of a fringing reef around its shores. There are many examples like this in the South Pacific. As such an island slowly subsides, in the manner shown in Stage 2 above, the surrounding reef gradually grows upward so as

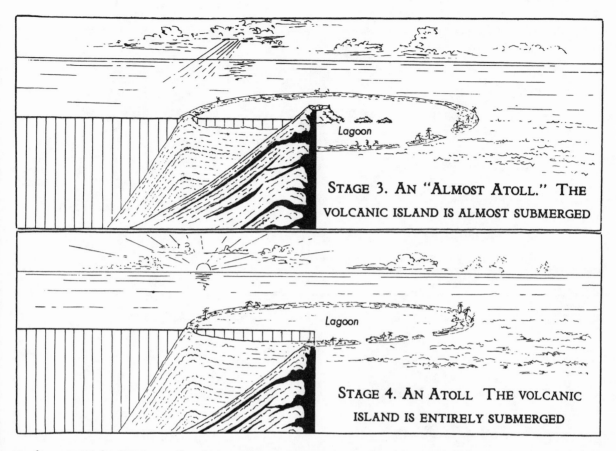

STAGE 3. AN "ALMOST ATOLL." THE VOLCANIC ISLAND IS ALMOST SUBMERGED

STAGE 4. AN ATOLL THE VOLCANIC ISLAND IS ENTIRELY SUBMERGED

to form a circular barrier reef enclosing a lagoon of shallow water. The coral reef grows upward in this way because the coral polyp thrives only in the agitated water on the outside of the reef. In the lagoon the polyp animals are choked by the silt which is washed into the lagoon from the island. Moreover, in quiet water the coral polyps do not grow so vigorously.

Further submergence of the island results in Stage 3. Here, the mere tip of the mountain remains above water, and the circular reef is far offshore. The atoll stage has almost been reached.

Finally, with complete submergence, an atoll. results. The atoll is almost never a perfect ring, but is broken here and there by passages through which ships may enter the quiet waters of the lagoon. In the course of time the reef gains a slight elevation above sea level. The wind may form dunes from the sand of the ground-up coral exposed at low tide. Vegetation in time develops, plants being carried here by ocean currents from other, similar islands or from distant mainlands. The cocoanut palm of course is by far the most common plant in this environment. This is largely because of the fact that the cocoanut seed is able to withstand long voyages in salty water as it is being transported by ocean currents.

Many atolls have become settled, and in recent years have served as landing fields on long ocean flights. Canton Island, almost on the equator, is perhaps the best known by present-day voyagers. Here the big planes en route to Australia from the United States and Canada come down in the cool of midnight to refuel for the next leg of their journey. Passengers with an hour or so on their hands can walk along the beach nearby, and by the light of the stars watch the many scuttling hermit crabs dodging the surf.

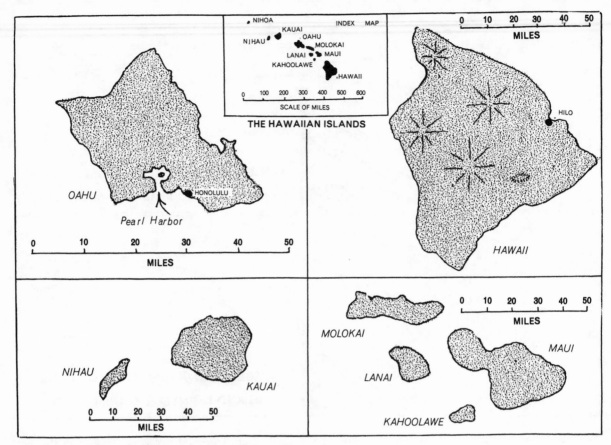

THE HAWAIIAN ISLANDS

ISLANDS. *Oceanic Islands. The Hawaiian Islands.*

    This well known chain of islands spans a distance of 1,800 miles in the central Pacific, from the big island of Hawaii at the east to tiny Midway Island at the far western end. This distance is, all told, almost as far as across the Atlantic. Only the eastern part of this long archipelago consists of large islands, the western members being mere reefs and atolls. There are nine islands which make up the eastern group, forming a chain about 500 miles long, as shown on the small insert map above. Eight of them are shown also on a larger scale, as they appear on many atlas maps.

    The most noteworthy feature about these larger members of the Hawaiian group is their great diversity in shape, as well as in size. It is hard indeed to believe that they could all have been formed in the same way. We have already noted, in Example No. 34, that many oceanic islands have had a volcanic origin. We are also familiar with the fact that the Hawaiian Islands support volcanoes that are still active. Of the 8 islands shown on the above map only Kauai, and perhaps Lanai, suggests a volcano by its outline. Maui, indeed, looks like two volcanoes fastened together. The other islands apparently are quite different in character. Molokai is much too long to be a volcano. The island of Hawaii seems much too irregular in outline to be a single volcano, and the island of Oahu is far too angular and straight-sided. The problem, then, is to account for all these different shapes and sizes. If these islands are volcanic in origin, and it is to be presumed that they are, then why are they so dissimilar in shape, and why also are their outlines so rarely circular? As a matter of fact, the Hawaiian Islands are unique, and differ from almost all other types of volcanoes to be found in the world.

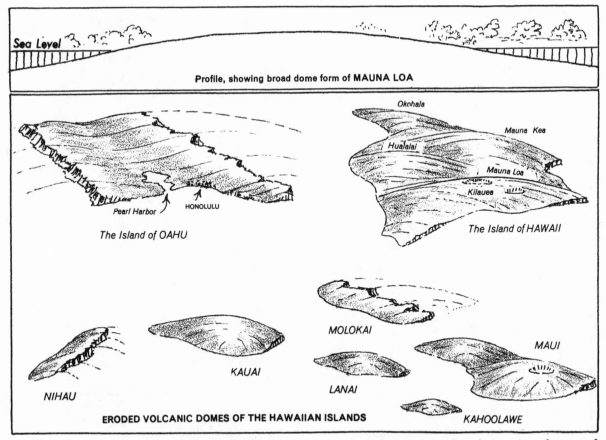

Profile, showing broad dome form of MAUNA LOA

ERODED VOLCANIC DOMES OF THE HAWAIIAN ISLANDS

There are two important facts about the Hawaiian Islands which help to explain their various shapes and sizes. The first has to do with the relatively broad dome-like form of the volcanoes themselves. The second is that several of the Hawaiian Islands have been almost entirely eaten away by the sea.

Concerning the dome-like form of the Hawaiian volcanoes, note, for example, the profile, in the drawing above, of the great volcano of Mauna Loa on the island of Hawaii. This section goes far below sea level, and indicates clearly the plateau-like character of this volcano. As the traveler sees this from the air, or from off at sea, he is usually surprised to observe the relatively flat top of this great mountain, which actually rises over 13,000 feet above sea level, and as much more above its submarine base. The famous crater of Kilauea, which has been made easily accessible by road, is simply a great steep-walled depression on the flanks of the volcano. There are several other similar craters.

Volcanoes of the Hawaiian type are formed by relatively quiet eruptions of lava which, pouring over the edges of the several craters, flow in broad sheets down the mountainside. In this manner the broad dome is gradually built up. This method of growth is quite in contrast with that of the tall conical peaks like Fuji Yama and Vesuvius and numerous other volcanoes throughout the world which were formed by explosive activity that ejected vast quantities of debris high into the sky.

Lying in the belt of the trade winds, the Hawaiian volcanoes have been subject to severe and incessant pounding by heavy seas. Some of the domes have been almost completely worn away, like Molokai, for instance. Others, like Oahu, upon which Honolulu stands, represent the remnants of two domes. In the depression between the two domes of Oahu lies the embayment which forms Pearl Harbor.

EXAMPLE 36

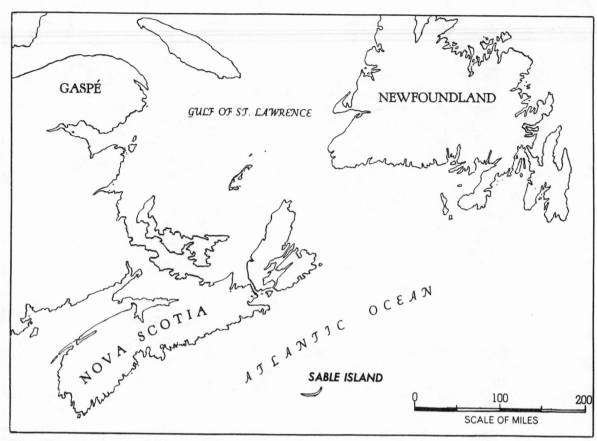

ISLANDS. *Isolated Islands. Sable Island off Nova Scotia.*

One hundred and ten miles off the coast of Nova Scotia, far out in the Atlantic, is a remote and strange island. It is unusual because of its distance from the mainland and its isolation from any other islands. It is the visible part of an extensive sand shoal which forms the great fishing "banks" of the Eastern Seaboard. These banks reach from Georges Bank in the Gulf of Maine all the way to Newfoundland, where they attain their greatest extent. On the banks themselves the water is quite shallow, being in some places less than 100 feet in depth. It is not shallow enough to impede navigation, except in the region of Sable Island, where countless shipwrecks have occurred. Closer inland the water deepens appreciably, reaching depths of several hundreds of feet.

Sable Island is 20 to 30 miles long and 1 to 2 miles in breadth. Its shape and size are constantly changing. In the days of the early navigators it was estimated to have been 100 miles in length, when it was known as Santa Cruz. Since 1763, when it was taken over by Britain, the island has diminished in size from 40 miles in length to 20, and from 2½ in breadth to about 1; and from 200 feet in height to 85. This is the elevation of the present sand dunes which cover its inland part. Large fresh-water lakes have been present at various times, and fresh water is always available. Wild ducks, gulls, and other birds nest there in large numbers, and wild fruits grow plentifully. There is a lighthouse at each end of the island, and a few other families live there. Cranberries are raised for export, and offshore there is a plentiful supply of fish. Because of the shifting shore line, which has at several times necessitated the moving of the lighthouses, with consequent menace to navigation, the Government has planted large quantities of trees and root-binding grass.

Similar to the banks of Nova Scotia and Newfoundland is the Dogger Bank in the North Sea, where the water is less than 80 feet deep. In fact almost anywhere in the North Sea you could put one of England's great cathedrals and half of it would be standing out of water. It is easy to understand, therefore, why it is possible to build a chain of man-made islands, the so-called Texas towers, on the banks off the Atlantic coast of the United States.

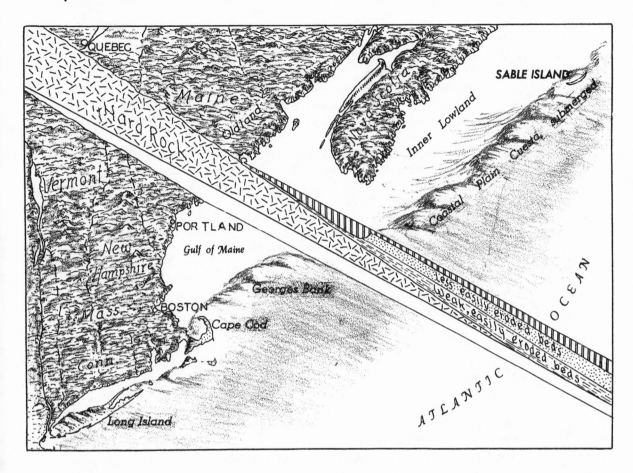

While Sable Island is unusual in being so isolated, it is, however, not unique. An island that is almost precisely like Sable Island in origin is Long Island, New York. The chief difference is that Long Island is much larger and is very close to the mainland. Martha's Vineyard might be cited as another example. It is about the same length as Sable Island but otherwise is much larger and, like Long Island, is much closer to the mainland.

These three islands, Sable Island, Martha's Vineyard, and Long Island, are all parts of the crest of the Coastal Plain cuesta. The above diagram is an attempt to show what this cuesta is like. A cuesta is a plateau-like part of a coastal plain which is separated from the so-called "oldland," upon which the coastal plain rests, by a lower belt or "inner lowland." This lower belt has been eroded out along the outcropping edges of weaker formations, as shown in the cross section above. In New Jersey this inner lowland is dry land and coincides with the narrow waistline crossing the central part of the state. But north of this point the lowland is submerged because of the lowering of the coast, and forms Long Island Sound and the deeper inshore water of the Gulf of Maine.

New England and Nova Scotia represent the "oldland," that is to say the hard-rock continental mass upon which the Coastal Plain was deposited. The Coastal Plain, after having been deposited, was raised above sea level and eroded by river systems. The inner lowland portion was eroded lower than the cuesta portion because here the underlying weak formations came to the surface. After the erosion of the inner lowland and the formation of the cuesta, the region was depressed and the inner lowland became submerged or drowned. In only a few places does the crest of the cuesta appear above sea level; namely, on Long Island, Martha's Vineyard, on Cape Cod, and on remote Sable Island. Elsewhere the cuesta lies under shallow water to form Georges Bank and the other well known fishing banks of the Atlantic Seaboard.

RIVERS

Before proceeding with our first example of rivers on the next page, it will be profitable to introduce this part of our subject with some general remarks.

Up to this point in our treatment of shore lines and islands we have been concerned with the features which form the margins of the continents. It is these features which are most prominently displayed on maps. They are the ones which we are sure to notice and about which we raise questions. The interiors of the continents are of course just as diverse as are the margins, but on most maps about the only clue we have to these diverse features is the rivers and lakes. On most maps rivers and lakes are likely to be obscured by all kinds of information—boundary lines of all sorts, cities, railroads, roads, elevation tints, and usually a profusion of names. Some maps are virtually clogged up with lettering. For these reasons it will be necessary for us to use care in selecting maps which portray enough rivers, mainly the smaller rivers, to provide interesting problems for us. If, instead of the usual small-scale atlas and road maps, we were supplied with a collection of the splendid topographic maps now issued by all the major countries of the world, we should have at hand many problems in map interpretation demanding a solution. It might be necessary in that event to have at our command a fair amount of basic geological information, if we were to try to "read between the lines" and attempt to infer a lot of facts which the map itself does not directly convey.

The problems and their explanations which we are about to set before you will, however, be simple ones. Nevertheless some information about the behavior of rivers will prove useful, and we shall therefore proceed to present it as follows:

1. The first fundamental fact about rivers, insofar as it may assist in the appreciation of the location and pattern of rivers on a map, is that in the long run rivers adjust themselves to the rock structure of the country. They tend in time to locate themselves upon the weaker, that is, the more easily eroded, formations. The river pattern therefore eventually comes to resemble the pattern of the rock outcrops. Inasmuch as certain types of rock structure have certain characteristic patterns of rock outcrop, river patterns will serve as a clue to these rock structures.

2. Among the characteristic patterns of rivers that will be illustrated are *parallel* streams; *trellis* drainage pattern; *rectangular* drainage pattern; *radial* drainage pattern; and the *annular*, or ring-like, drainage pattern. There are also streams with *barbed* tributaries, as well as other varieties. Among our atlases and road maps we shall be able to find examples of each.

3. A third important fact about streams is that they encroach upon new territory by working headward after the manner of gullies on a hillside. Usually rivers work headward along the least resistant paths by eating upstream along the weaker rock outcrops. As a consequence, streams may encroach upon the courses of other streams, and divert them by capturing their upstream or headwater portions. Stream capture or piracy accounts for many of the right-angled bends which we find in rivers, many indeed which we may run across in our perusal of maps.

4. Then again we have the interesting fact that rivers can not always choose the courses which they would find easiest to follow. Some external agency, like a continental ice sheet, may entirely obliterate or disturb a drainage system. When in that case the ice disappears, the new drainage system which again occupies the country will not resemble at all the previous one, and will show no adjustment to the rock structure.

5. Then there are some situations which are real puzzlers. We find a river like the Danube, for example, flowing for many miles over a vast flat plain like that of Hungary, and then suddenly coming to a mountain barrier which it breaks through in a narrow gorge, as at the Iron Gates across the Transylvanian Alps. How, in the normal course of development, could a river ever acquire a course like that? There are in the world many water gaps like the Iron Gates. Some of these we shall try to explain.

6. Finally, as we come to the subject of lakes, we shall have to remember that many lakes are nothing more nor less than rivers that have been blocked by some kind of dam. Lakes in all their diversity have almost nothing in common except that they are bodies of water. As in the case of rivers, their location and shape may be attributed to almost every kind of geological event.

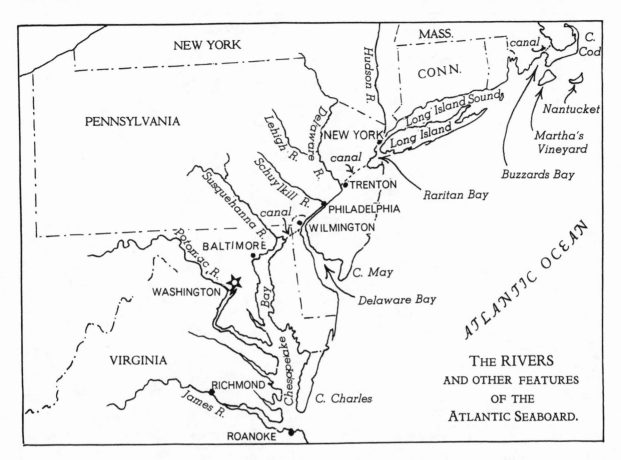

THE RIVERS
AND OTHER FEATURES
OF THE
ATLANTIC SEABOARD.

RIVERS. *River Embayments. The Delaware (Delaware Bay); the Susquehanna (Chesapeake Bay), and the Potomac.*

An examination of the above map shows a rather noteworthy fact about these three river embayments. They are similar in several respects. These rivers are alike in flowing first to the southeast. Then, as they enter their embayed portions, they turn abruptly to the southwest, to be followed shortly by another right-angled turn to the southeast again. Not only this, but we observe the further remarkable fact that the continuation of these embayments to the northeast falls in line with several other embayed water bodies; namely, Raritan Bay, Long Island Sound, and Buzzards Bay. Still further, there are three man-made waterways which fall along this same line and connect several of the embayments with each other. Between upper Delaware Bay and upper Chesapeake Bay is the important Delaware and Chesapeake Canal, which is used by large ocean-going vessels. Between the uppermost end of Delaware Bay, at Trenton, and Raritan Bay, there is the Delaware and Raritan Canal, once an important waterway but now in disuse. Between Buzzards Bay and Cape Cod Bay is the Cape Cod Canal, also used by large vessels. This alignment of water features (nine of them) along the Atlantic Seaboard surely requires an explanation.

Two or three other things may also be noted. Note that Martha's Vineyard Island lies just off the upper arm of Cape Cod. This may not seem noteworthy, but the reason for this will soon become apparent. More remarkable are the peninsulas of Cape May and Cape Charles. These two peninsulas almost duplicate each other, and so also do the general shapes of Delaware and Chesapeake bays.

The explanation for all of these things is a fascinating story which, with the help of maps, will be treated on the two following pages. The explanation will probably raise more questions than it answers, but that is the way with most scientific discoveries.

EXAMPLE 37

The Explanation

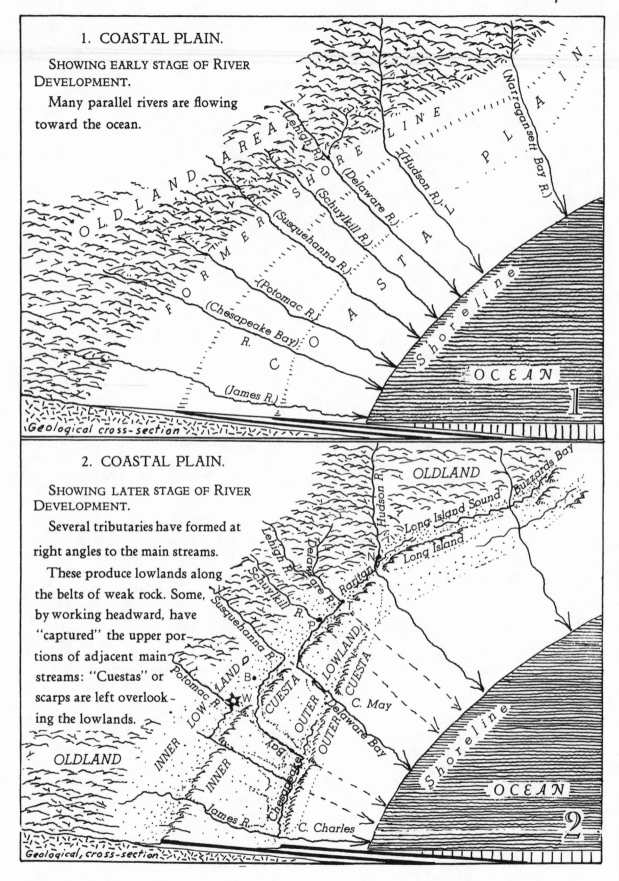

1. COASTAL PLAIN.

SHOWING EARLY STAGE OF RIVER DEVELOPMENT.

Many parallel rivers are flowing toward the ocean.

OLDLAND AREA

FORMER COASTAL SHORE LINE PLAIN

(Lehigh R.)
(Schuylkill R.)
(Delaware R.)
(Hudson R.)
(Narragansett Bay R.)
(Susquehanna R.)
(Potomac R.)
(Chesapeake Bay)
(James R.)

Shoreline

OCEAN

Geological cross-section

2. COASTAL PLAIN.

SHOWING LATER STAGE OF RIVER DEVELOPMENT.

Several tributaries have formed at right angles to the main streams.

These produce lowlands along the belts of weak rock. Some, by working headward, have "captured" the upper portions of adjacent main streams: "Cuestas" or scarps are left overlooking the lowlands.

OLDLAND

Hudson R.
Long Island Sound
Buzzards Bay
Long Island
Raritan
Lehigh R.
Schuylkill R.
Delaware R.
Susquehanna R.
Potomac R.
LOWLAND
INNER LOWLAND
OUTER CUESTA
INNER CUESTA
Chesapeake Bay
Delaware Bay
C. May
C. Charles
James R.
OLDLAND

Shoreline

OCEAN

Geological, cross-section

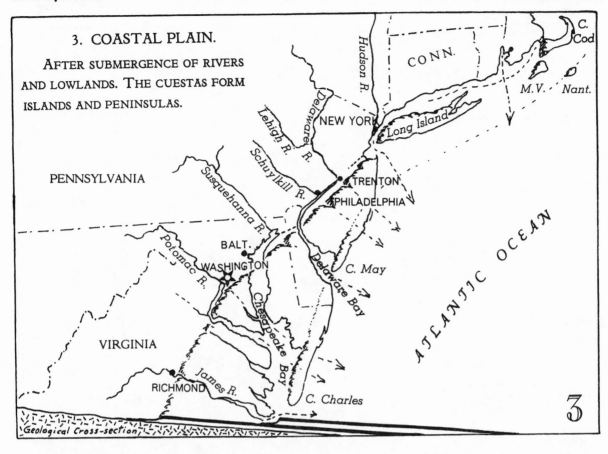

3. COASTAL PLAIN.

AFTER SUBMERGENCE OF RIVERS AND LOWLANDS. THE CUESTAS FORM ISLANDS AND PENINSULAS.

The three diagrammatic maps on these two pages show very simply the essential steps which have brought about the features of the Atlantic Seaboard discussed on the previous page.

On Map No. 1, opposite, is shown a simple Coastal Plain recently elevated from beneath the ocean. Because the Coastal Plain is the former sea floor, it is comparatively flat, and slopes off gently seaward from the "Oldland" area. Flowing across this Coastal Plain are several streams, eight of them, named so that we can keep track of them and later identify them. Starting at the south, they are the James, Chesapeake Bay R., Potomac, Susquehanna, Schuylkill, Delaware, Hudson, and Narragansett Bay R. Another small tributary is the Lehigh. In the front of this map is a simple geological cross section. This shows the layers which make up the Coastal Plain, dipping or sloping gently toward the ocean.

Map No. 2 shows a later stage. The rivers are now eroding their valleys below the surface of the plain. Particularly noteworthy is the fact that several tributaries to these main streams have developed, and enter the main stream at right angles. In several cases these tributaries have eaten their way headward along the belts of weaker rock and have captured and thus diverted the upper portions of some of the adjacent main streams. The weak beds, usually clays, are the light ones on the cross section; and the two resistant beds are the two heavy lines. The two resistant beds remain uneroded to form scarps or cuestas which face away from the ocean and look down upon the lowland belts. This unique process of development, very briefly suggested here, is typical of Coastal-Plain development the world over.

In the final stage, Map No. 3, above, we see the present-day features of the region. This condition has been brought about by a slight lowering of the area and the "drowning" or submergence of the lower ends of the river valleys, together with parts of the two lowland belts. The inner cuesta remains to form such features as the arm of Cape Cod, Martha's Vineyard Island, Long Island, and the belt of hills crossing New Jersey and extending southward. Parts of the outer cuesta form Cape May and Cape Charles, as well as Nantucket.

EXAMPLE 38   *The Problem*

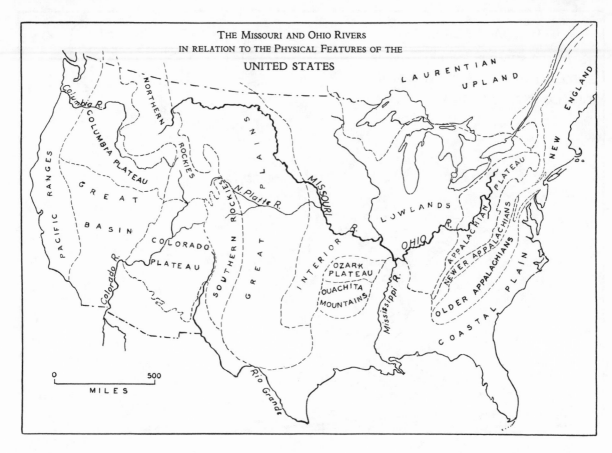

THE MISSOURI AND OHIO RIVERS
IN RELATION TO THE PHYSICAL FEATURES OF THE
UNITED STATES

RIVERS. *River Locations. The Missouri and the Ohio.*

The great rivers of the world, some of them flowing for thousands of miles across a continent, appear to be fixed and permanent features of the landscape. This is true only in a relative way. Some rivers have existed doubtless for untold ages. Others, and large ones at that, have come into being only recently, almost within the memory of man. Of course, numerous small changes in the flow of rivers is always going on. But our interest in this example is in two great rivers which have come into being full blown where no river existed before.

The Missouri and the Ohio rivers, flowing across almost unobstructed country, appear to follow courses taken almost by chance. The Missouri River rises in the northern Rockies and finds its way across the Great Plains and the equally great Interior Lowlands of the United States to the Mississippi, in all a distance of more than 1,500 miles. The Ohio, with its northern tributary the Allegheny, has half the length of the Missouri, and like that stream traverses two of the physiographic provinces of the country, the Appalachian Plateau and the Interior Lowlands. Both of these streams encounter small waterfalls and rapids along their courses, but neither one apparently is obliged to circumvent any mountainous barrier. Examples of mountain barriers which deflect rivers may be seen in the great mass of the northern Rockies around whose southern base the Snake River flows for hundreds of miles, and also the northern tip of the southern Rockies, which guides the course of the North Platte River.

Most rivers have quite complex histories, the various segments having originated in various ways, and then later assembled to make a single unit. The Missouri and the Ohio, however, have been guided throughout their courses by one controlling event, which is described on the next page.

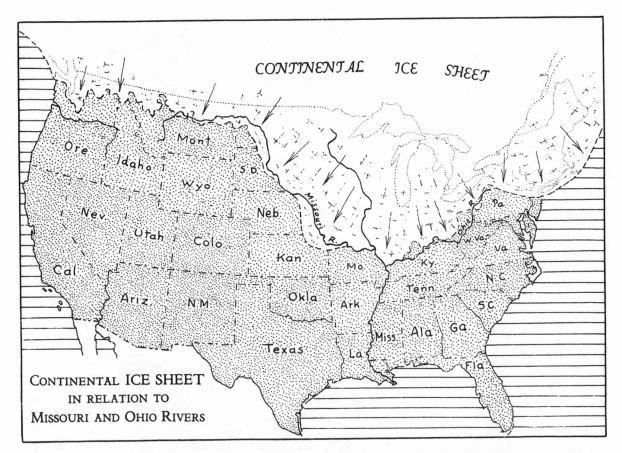

CONTINENTAL ICE SHEET
IN RELATION TO
MISSOURI AND OHIO RIVERS

The Missouri and the Ohio rivers are located where they are because of the continental ice sheet. The above map shows the continental ice sheet at the time of its most southerly advance. The margin of the ice coincides roughly with the position of the Missouri and Ohio rivers.

In order to understand how these river changes came about, it will be necessary to try to visualize just what was happening along the ice front at this time. First of all, were we to observe the ice front over a period of some few thousands of years we should note that its position was constantly changing. From season to season this change might be negligible. Even during a person's life the change might be only a few hundreds of feet, probably less than a mile or so. But over the centuries the ice would be seen to melt back many miles, and at other times to advance again. Over still larger intervals of time, measured in the tens of thousands of years, the ice front might recede even back to Canada, leaving the United States uncovered for a long interglacial period, this to be followed by a later advance. Because of this behavior several glacial stages are recognized, mainly from the glacial deposits which were left. With its interglacial stages the whole Glacial Period embraced one to two million years. Thus we see it was a continual problem on the part of the streams to maintain their courses.

Streams from the Rocky Mountains, flowing eastward to the Mississippi, were time and again blocked by the ice and forced to work their way along the ice front. Streams flowing northward to the Great Lakes from the Ohio River Basin were likewise obstructed and forced to seek out transverse channels along the ice margin. These many channel ways, all more or less parallel with the ice front, constitute today the main courses of the Missouri and Ohio rivers. Waterfalls, as is to be expected, interrupt these courses at many points, notably at Great Falls, Montana, on the Missouri, and also at Louisville on the Ohio. In some places, particularly in Ohio, the old south-north river valleys, now buried under glacial drift, have been detected in well borings.

EXAMPLE 39   *The Problem*

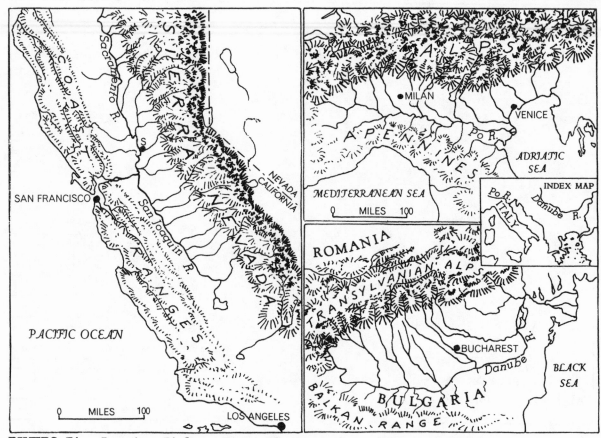

RIVERS. *River Locations. Piedmont Rivers. The San Joaquin, the Po, and the Danube Rivers.*

There are depicted upon the above maps three well known rivers, the San Joaquin, the Po, and the Danube. They are all located in broad valleys or basins at the foot of large mountain ranges. These are only a few of the many examples of such rivers that are to be found throughout the world.

Notice that in all of these cases the rivers do not flow close to the base of the larger mountains. They seem actually trying to get as far away from the high mountains as they can.

There is another noteworthy fact also. Observe that many large streams enter each of these main rivers from one side and that very few enter them from the other side. The more numerous and the larger tributaries in each case come down from the higher mountains. Let us now examine each main river in detail.

The San Joaquin flows along the western side of the Great California Valley, close to the smaller Coast Ranges and far from the foot of the mighty Sierra Nevada. Virtually no streams enter the San Joaquin from the side of the Coast Ranges.

The Po River of the northern Italian Piedmont flows close to the base of the relatively low Apennines and far away from the base of the Alps. From the Alps, also, it receives all of its larger tributaries.

The Danube, likewise, lies close to the Balkan Range and far from the much larger Transylvanian Alps on the north. It also receives its largest tributaries from the north.

Another striking example of this kind is the Ganges in India. It lies far south of the base of the Himalayas, from which also it receives most of its tributaries. The Garonne in southern France, the upper Danube in southern Germany, and the Paraná River in northern Argentina are other rivers exhibiting this behavior, that can be found in any atlas.

HIGH MOUNTAINS, like the Sierra Nevada and Alps

A L L U V I A L   F A N S

P I E D M O N T   R I V E R, like the   S a n   J o a q u i n   and   Po

LOW MOUNTAINS like the Coast Ranges and Apennines

The accompanying sketch is designed to show why rivers tend to be located far from the base of high mountain ranges. They are actually pushed over there by their tributaries.

High glaciated mountains like the Sierra Nevada, the Alps, the Transylvanian Alps, the Himalayas, and the Pyrenees usually have large streams which carry a great deal of detritus. During the relatively recent Glacial Period, when the glaciers were larger and more numerous than they are at present, the rivers emerging from the melting glaciers were heavily laden with sand and gravel. This load of material was carried down to the foot of the mountains and was there deposited in the piedmont basins in the form of outwash plains or alluvial fans. As the fans spread outward into the basin they forced the main river draining the basin farther and farther away from the foot of the high mountain range.

The smaller mountains, like the Coast Ranges and the Apennines, at whose foot the main river was eventually forced to flow, were never glaciated. Their smaller rivers, therefore, never carried any great quantities of silt into the piedmont lowland. Because of this, no alluvial fans were built at the base of the lower ranges.

In some atlases, as for example *Bartholomew's Advanced Atlas of Modern Geography* which I have before me, there are occasional geological maps of the continents, or of the whole world. In all the basins that have been mentioned; namely, those drained by the San Joaquin, the Po, the Danube, the Ganges, and the Garonne, you can note on the geological map the fact that these basins are filled with what are called "Quaternary" deposits. This means alluvial formations of Glacial Time. Even now these streams are carrying heavy amounts of detritus. This is evident to visitors, who are certain to remark on their milky appearance, caused by the suspended clay particles. Such streams are often called "glacial milk."

EXAMPLE 40   *The Problem*

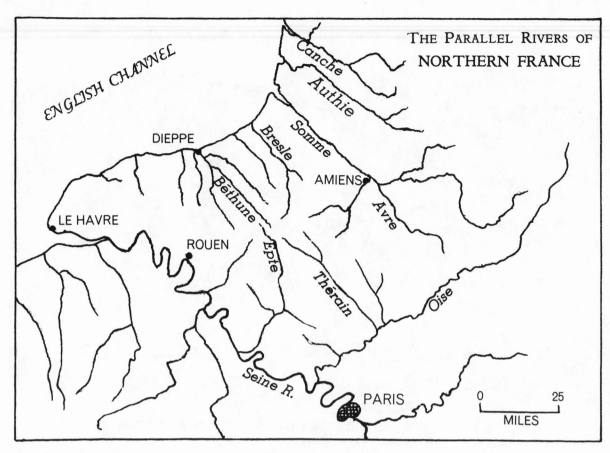

THE PARALLEL RIVERS OF
NORTHERN FRANCE

ENGLISH CHANNEL

DIEPPE

LE HAVRE

ROUEN

Canche

Authie

Somme

Bresle

AMIENS

Bethune

Avre

Epte

Thérain

Oise

Seine R.

PARIS

0   25
MILES

RIVERS. *River Patterns. Parallel Rivers. The Rivers of Northern France.*

Although most of the smaller rivers of the world seem to flow in a hit-and-miss fashion, wandering at will with little apparent plan, nevertheless a careful inspection of maps shows that this is not true in all cases. Many rivers flow in definite patterns. We must assume that there is a cause for this behavior, if we can but find it.

Perhaps the simplest pattern of all is that in which streams flow parallel with each other. Even this simple pattern, however, may be brought about by different causes. In this example and the succeeding one, two different categories of parallel streams are considered.

One of the rather unusual examples of parallel streams is in northern France, north of Paris. A simple sketch map of this area is given above. This map shows not only about a dozen streams which flow parallel to one another toward the English Channel, but also several other streams which have a similar alignment but which flow in the opposite direction. The Seine is by far the largest of all these streams. In spite of its many meanders it maintains a fairly direct course to the Channel. It is interesting to note also that many of the small tributaries to the major streams flow in a similar direction.

Obviously this whole situation cannot be passed over as being purely accidental. There is definitely some controlling reason. Once determined, the explanation for the parallelism of these rivers may possibly be also the explanation for certain other features of the landscape, and may perhaps even be fundamental to an understanding of this entire area.

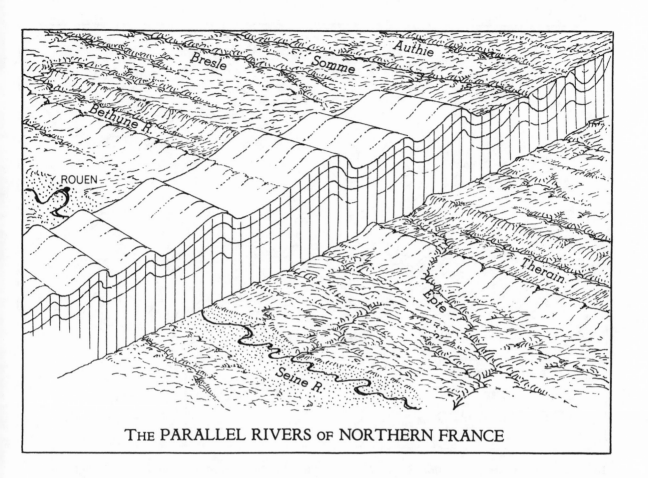

THE PARALLEL RIVERS OF NORTHERN FRANCE

The parallel rivers of northern France are controlled in their position by the parallel folds or corrugations of the underlying rock structure. This is illustrated in the above diagram. It is to be noted that the original folds are no longer intact. They have been beveled off to form an almost level plain. Even so, the rock outcrops have a parallel trend, and it is these belts of alternating resistant and weak formations which determine the stream positions.

The middle belt of the drawing above represents the original folds of the region. Sedimentary beds of several kinds are included in the folding. In northern France these beds are mainly shales and clays, limestones, and chalk. The shales and clays are weak, that is to say, easily eroded. The chalk, strangely enough, is resistant. It therefore forms valley walls and scarps.

The largest of the folds, the one appearing in the middle of the drawing, has been worn down so as to expose along its axis or central line the weaker underlying clays of the region. It is along this clay belt that the Béthune and the Thérain rivers have developed their courses, one to the northwest, the other to the southeast. The elongated valley occupied by these two streams is known as the Pays de Bray. Having a length of some 50 miles, and hemmed in by chalk scarps 600 to 700 feet high, it is a little world by itself. It is a veritable hothouse protected from the raw winds of the upland above. As a result it supports many fruit and dairy farms. Its market center, Neufchâtel, is noted for its cheeses.

EXAMPLE 41   *The Problem*

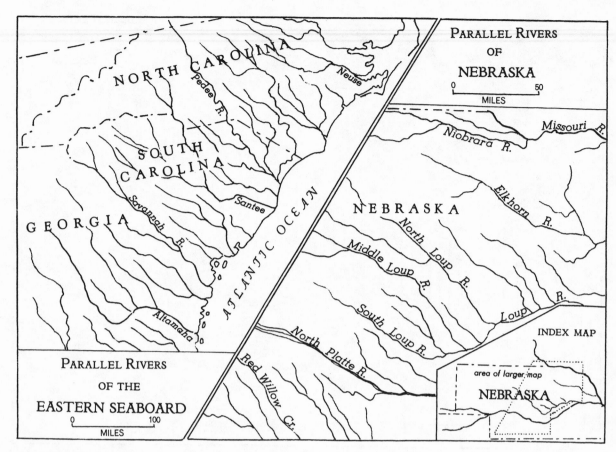

RIVERS. *River Patterns. Parallel Rivers. Rivers of the Atlantic Seaboard; Some Rivers of Nebraska.*

Parallel rivers occur in only a few parts of the world. There are of course many places where two or even three streams are parallel, but to find dozens of them close together, all flowing more or less in the same direction, is unusual. The rivers of Virginia, North and South Carolina, and Georgia illustrate this admirably. Not only do the main streams flow parallel, but the tributaries also flow many miles alongside each other before joining their master stream. Notable also is the fact that there are almost no small tributaries coming in at right angles to the main streams. There is in fact hardly any room or need for these additional drainage lines. If for any reason any stream departs from its direct course to the sea, it soon turns again and regains its former direction. There is obviously some strong compelling force which influences all of these rivers.

In part of Nebraska the parallel river systems are almost as crowded as they are in the Carolinas. The Nebraska portion of the above map, because of its larger scale, represents about half as great an area as that covered by the other map. The density of streams, therefore, is about the same in the two cases. One of these areas is near the ocean; the other is more than a thousand miles inland; yet they have very similar characteristics.

Glance, if you will, at some other maps in your atlas, and you will see that you can hardly duplicate these examples. Europe and Africa provide none. Possibly the vast lowlands of northern Siberia do, but even there the rivers are hardly parallel nor is there such a profusion of smaller streams. The region around James Bay in Canada is not too bad a case, nor indeed also are certain portions of South America. It would seem, therefore, that the parallel pattern, simple as it appears to be, is not nearly so common as one might at first imagine.

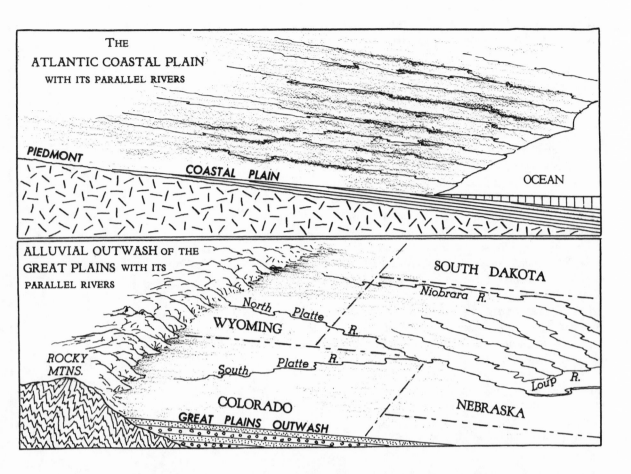

Parallel rivers develop on those parts of the earth's surface where the land is relatively flat, but slopes off gently in one direction. Such an area resembles a slightly tipping roof, the kind that one might find on a factory building. Such a roof slants just enough to carry off the water. Rain falling on such a roof runs off in a sheet or, what is more likely at the beginning of a rainstorm, in a number of parallel rivulets.

The Atlantic Coastal Plain is an almost perfect example of such a surface. This plain slopes imperceptibly toward the sea. It is made up of strata that are almost as regular as sheets of cardboard lying flatwise one on top of another. As the coastal plain was gradually uplifted above the ocean, streams formed upon its surface. Doubtless, also, at one time the coastal plain extended much farther inland than it does now. Its thin inner portion, after having served to determine the direction of stream flow, has been eroded away, leaving the slightly rolling topography of the Piedmont. Here the once-buried old crystalline or granitic rocks are now brought to view.

In certain parts of the world there are, at the foot of mountain ranges, great outwash deposits or alluvial fans. These, as the name indicates, are usually fan-shaped, being formed by a large stream that emerges from the mountains and breaks up into a number of distributaries. However, as in the case of the Great Plains of the United States, numerous streams for a great distance along the front of the Rocky Mountains carried out extensive sheets of alluvium, that is to say, gravel and sand. This now forms a continuous plain which slopes away almost uniformly from the mountains. The deposits making up the Great Plains reach 500 miles to the east, becoming thinner and thinner in eastern Nebraska and the Missouri River region. The sketches above have been rendered quite diagrammatic in order to emphasize the conditions just described.

EXAMPLE 42

The Problem

RIVERS. *River Patterns. More Parallel Rivers. Yazoo Type. The Yazoo and Mississippi Rivers.*

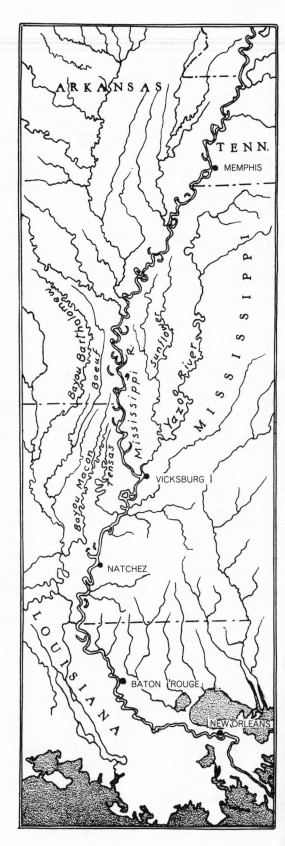

An examination of almost any map of the lower Mississippi system shows a remarkable tendency of the tributary streams to flow for many miles parallel to the Mississippi before joining it. The Yazoo is but one of many examples. Tributaries to the tributaries behave in the same way. Look at maps of other rivers, such as the Ohio and the Allegheny, and great streams like the Colorado and the Columbia rivers, and note that their tributaries join them more nearly at right angles. On the other hand streams like the Indus and the Ganges systems, the Tigris and the Euphrates, and also the lower Hoang Ho (Yellow River) all show characteristics like that of the Mississippi.

As we puzzle over this peculiar behavior of the tributaries of so many of the large rivers of the world, and examine the various atlas maps, we note that these streams, in their lower courses, where their tributaries flow parallel to them, all occupy extensive flat areas of country. In their upper courses, where the country is more rugged and mountainous, their tributaries behave in the usual way, that is they are apt to be dendritic and irregular. It would appear, therefore, that where the land is flat streams tend to flow parallel to one another.

The lower Mississippi, say from Memphis, Tennessee, south to the Gulf of Mexico, occupies an extensive flood plain, a wide flat-floored valley, along which it pursues its meandering course. This is the part of the river made famous by Mark Twain in his *Life on the Mississippi*. This flood plain is a dozen miles or more wide, with bluffs on either side. Only occasionally along its course does the Mississippi impinge against the bluffs, as it does at Vicksburg, Natchez, and at Baton Rouge. In times of flood the tributaries themselves may disappear, the whole area becoming a vast lake.

At the southern end the flood plain merges with the Mississippi delta. In fact, from Memphis southward, the entire flood plain is essentially the delta which began to form many years ago at the head of a great embayment and which has since been steadily extending south, even out into the Gulf.

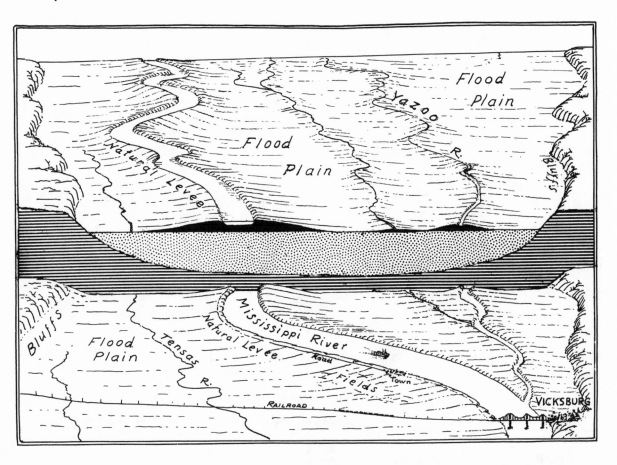

The relationship between the Yazoo River and the Mississippi River is shown on the above diagram. Running through the middle of the diagram is a cross section. This cross section shows three kinds of formations. First is the bedrock of the country shown by the horizontal ruling. This is part of the coastal plain of the southern United States. These rock formations now form the bluffs on either side of the wide flood plain of the Mississippi. The flood-plain deposits are shown by the dot pattern. These deposits fill a broad trench-like depression originally cut by the Mississippi below the Coastal Plain beds.

Large rivers, like the Mississippi, carry great quantities of sediment. They are therefore muddy. At times of high water these large rivers overflow their banks. When they overflow their banks the speed of their flow is reduced because of the friction, and some of the muddy sediment is deposited. In other words, the bank of the river is thus built up or elevated. Such a ridge is called a "levee," a word derived from the same root as "elevate," to raise. With deposition taking place also to some extent on the bed of the river, the entire stream is raised and comes to flow on a levee ridge. This ridge may be 10 to 20 feet above the flood plain on either side.

Because of the elevated position of the river, the tributaries on the flood plain can not get into the main river. They therefore are forced to remain on the flood plain and flow parallel to the main river for many miles. In some cases the tributaries become blocked by the levees of the main river and form lakes, but usually the tributaries flow along the side of the flood plain until they reach some point farther downstream where the main stream swings against the valley walls. The Yazoo joins the Mississippi where it comes against the bluffs at Vicksburg. Some rivers, like the Atchafalaya, never actually get to be tributaries because they reach the ocean first.

EXAMPLE 43   *The Problem*

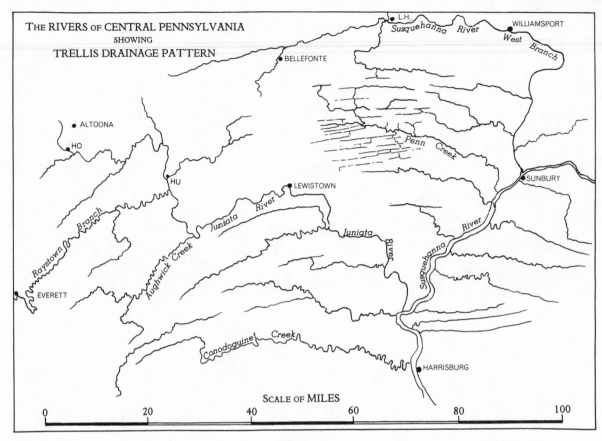

THE RIVERS OF CENTRAL PENNSYLVANIA
SHOWING
TRELLIS DRAINAGE PATTERN

SCALE OF MILES

0   20   40   60   80   100

RIVERS. *River Patterns. Trellis Rivers. Pennsylvania.*

In central Pennsylvania the main rivers and their tributaries exhibit a pattern which occurs only under certain special conditions of geological structure. The map above shows some of this area. This region of trellis drainage extends eastward into the states of New Jersey and New York, and southward into Maryland and West Viriginia, and then all the way to eastern Tennessee and Alabama. It is a long belt of country. The rivers shown on the above map are approximately the same as those shown on the automobile road maps of Pennsylvania, and also on the maps of the National Geographic Society. Only the larger streams are shown on these maps because they are relatively small-scale maps. Large-scale maps, like the United States Geological Survey topographic sheets, show many more of the smaller streams and tributaries. On a small part of the map above, the Penn Creek drainage area, these smaller streams have been added, as they are represented on the Millheim Quadrangle of the Survey topographic maps. The rest of the map would resemble this small area if the complete stream systems were shown.

It is easy to see, when we glance at this map, why this stream pattern is called "trellis." The streams branch like a grapevine growing on a trellis or arbor. All of the streams here belong to the Susquehanna System. These tributaries all join the Susquehanna above Harrisburg. The West Branch of the Susquehanna and of the Juniata rivers constitute the chief tributaries. In the small area where the detailed drainage pattern is shown, practically all the smaller tributaries belong to the Penn Creek System. All of the other streams have just as many small branches.

The trellis pattern of streams seems to be made up of streams which in part of their course run transverse to the grain of the country. On this map, that is more or less north and south. In other parts of their course they run longitudinally. Raystown Branch and Aughwick Creek, with its continuation the Juniata River, are longitudinal streams. They flow more or less east and west, which, as we shall see, is with the grain of the country. These longitudinal streams flow at right angles to the transverse ones.

90

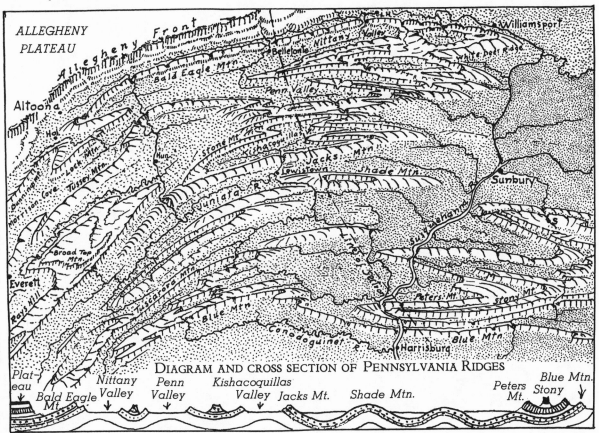

DIAGRAM AND CROSS SECTION OF PENNSYLVANIA RIDGES

The diagrammatic map above shows exactly the same area as that depicted on the preceding page. It represents diagrammatically the topography of the region. This part of Pennsylvania consists of many long hills or ridges of various shapes. These ridges are really much bigger than they might seem on this small sketch. They are, most of them, veritable mountains, being 1,000 to 2,000 feet in height. Mountains of this kind are brought about by folding of the earth's crust. The geological cross section at the bottom of the above sketch shows what this folding is actually like. Not only have the rock formations been folded but they have also been profoundly eroded. At the very eastern end and at the western end of the cross section appear several beds at the top of the series of beds, which have been completely worn away in the middle part of the cross section. As a result of erosion, therefore, several different kinds of mountain may be produced.

Take, for example, the three types of mountain represented by Shade Mountain, by Stony Mountain, and by Jacks Mountain. Shade Mountain is a complete uparched fold. It is called an "anticlinal mountain." Stony Mountain is just the opposite. It is actually the central part, that is, the axis, of a downwarped fold, or "syncline." Jacks Mountain, like Bald Eagle Mountain farther west, is just one side of a fold. The valleys are equally varied. The three named valleys in the cross section, that is, Nittany Valley, Penn Valley, and Kishacoquillas Valley, are all parts of original uparched folds whose entire tops have been worn off. The long broken line on the map shows the location of the cross section. The several features on the map and on the cross section can be identified with one another.

The zigzag shape of the mountains shown on the above diagram is one of the unique and rather surprising things about the Pennsylvania folded mountains. They are actually wrinkles of the earth's crust which have been beveled across, a condition which can readily be duplicated with a sheet of paper and a pair of shears. The Indians used to call these the "Endless Mountains." It is easy to see why.

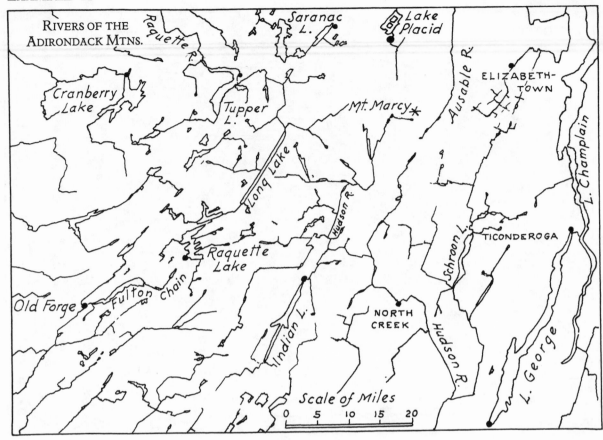

RIVERS. *River Patterns. Rectangular Rivers. The Adirondack Mountains.*

Another somewhat geometrical pattern of river system is the angular, or what is frequently a rectangular, pattern. It is true that the trellis pattern is also rectangular, but if we compare the previous map with the one above we see that there is quite a difference between the two.

Streams with a rectangular pattern turn around sharp corners, and their tributaries come in at right angles. Note particularly that there is not the close parallelism of streams such as occurs in those which have the trellis pattern.

The rectangular pattern is far more common than we are apt to think because it is not shown so frequently on small-scale maps. The reason for this is that the rectangular pattern is most commonly developed in the smaller tributaries rather than in the main streams.

The map above, showing the major streams of the central Adirondack Mountains of northern New York State, reveals numerous instances of rectangularity. If all the small tributaries were to be shown, as has been done above in the Elizabethtown area, the right-angled checkerboard-like pattern would be even more noticeable. The remarkable thing about this pattern is that almost no one ever notices it. The rivers represented on the above map are exactly those which appear on most of the automobile road maps. Inasmuch as the roads usually follow the watercourses, the roads also, especially the main ones, also have a right-angled pattern. In certain places the right-angled pattern of streams does not escape anyone's attention. In the Ausable Chasm, for instance, in the northern Adirondacks, those who go through this scenic gorge by boat are quite aware of the right-angled turns.

In the Adirondacks, as may clearly be seen on the above map, many of the lakes and watercourses trend in a northeast-southwest direction. In fact, the lines of this trend are carried through from one drainage system to the next with only occasional breaks. Other lines transverse to these are also common, and these lines likewise may be carried through from one part of the map to another, with only a few interruptions.

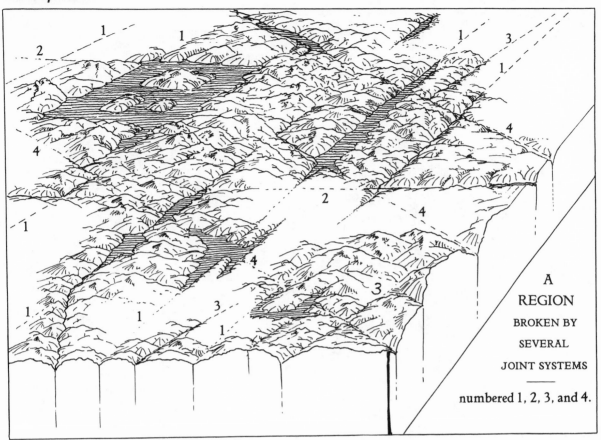

A
REGION
BROKEN BY
SEVERAL
JOINT SYSTEMS
———
numbered 1, 2, 3, and 4.

The explanation for a rectangular or an angular stream pattern is a simple one. This pattern is due to the fact that the region involved is broken by a series of crisscrossing fractures or joints (cracks). A diagrammatic sketch of such a region is shown above. Large rock masses like those which make up the Adirondacks are brittle; and when they are subjected to strains, as they are during earth movements, they break or shatter like any other brittle material. These breaks are usually at right angles to each other, as has often been demonstrated in laboratory experiments. These various sets of cracks are known as "joint systems." On the sketch above the joints marked "1" are intersected by a second set marked "2" which is transverse to the first. The diagonal joints marked "3" and "4" are also transverse to each other. There may be joints in other directions as well.

Cracks or joints are zones of weakness, and it is therefore along these zones that streams eventually come to have their positions. In other words, the stream pattern of a region is a fair indication of the joint pattern of the region. For that matter, the stream pattern is an indication of any belt of weakness, whether it be joints or less resistant layers of rock.

In some regions of the world the joints are widely spaced, perhaps a mile or two, or even more, apart. The resulting topography is therefore rather bold or "coarse," as in the Catskills. In other places the joints may be quite close together, many joints to a mile. The result then is that there are many small streams close together, and the topography is "fine-textured." This is the case in the hills of West Virginia, and indeed in parts of the Adirondacks too. Raquette Lake, Cranberry Lake, Saranac Lake, and Lake Placid all show many small details which reflect the fine texture of the drainage lines.

On the above diagram several of the actual Adirondack lakes may be recognized, such as Long Lake, Raquette Lake with its many branches, and Lake Placid with its rectangular form and rectangular islands.

Rectangular rivers are to be noted in many parts of the world, such as southern Sweden and Finland, parts of Ontario, and even in the tributaries of the Grand Canyon.

EXAMPLE 45   *The Problem*

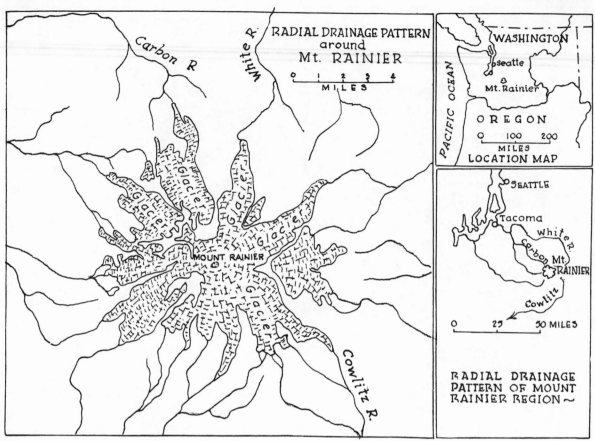

RADIAL DRAINAGE PATTERN
around
Mt. RAINIER

RADIAL DRAINAGE
PATTERN OF MOUNT
RAINIER REGION ～

RIVERS. *River Patterns. Radial Rivers. Mount Rainier.*

The examination of many atlases and other maps will often reveal certain symmetrically shaped mountains from which the streams radiate in all directions. There are actually several types of radial drainage patterns. The type to be considered here usually comprises only a relatively small extent of country. Most atlas maps, therefore, because of their small scale, do not depict all of the streams which make up the radial pattern. The maps, in short, are much simplified. Take, for example, the radial drainage pattern of Mount Rainier in the Cascades. The above small Location Map shows the position of Mount Rainier in the state of Washington. It is actually within sight of Tacoma and is in fact sometimes known as Mount Tacoma. Below the above Location Map is the representation of the drainage lines radiating from Mount Rainier as they appear on the National Geographic Society map of the Northwestern United States. The radial pattern is much simplified, and many of the smaller rivers are omitted.

In order to show more fully the complete drainage pattern of the area, we have reproduced above (but also somewhat simplified) the map which is made available to visitors in the small guidebook to Mount Rainier National Park. Almost two dozen streams are shown radiating from this great peak. Many of them are supplied by melting glaciers.

Several other peaks in the Cascades, such as Mount Baker, Mount Adams, and Mount Hood, exhibit similar radial drainage patterns. It should be noted that there are many other kinds of peaks throughout the world which do not display this kind of drainage design. The rivers flowing from Pikes Peak, for example, are by no means radial. All of this must mean that the peaks of the Cascades have had a unique origin, one that is different from that of many other famous mountains.

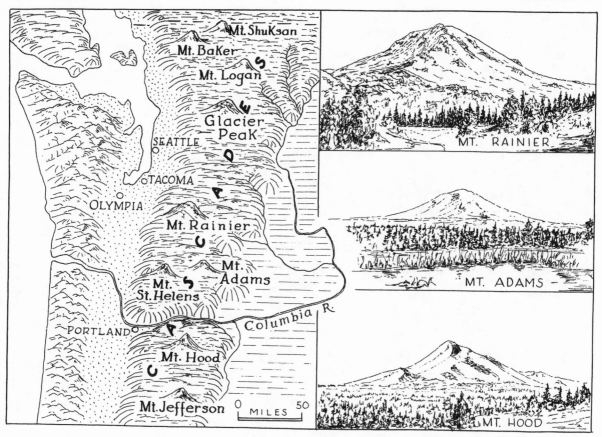

Mount Rainier and the other peaks of the Cascades stand like solitary sentinels above the plateau-like Cascade upland. All of them are volcanoes. Starting at the north near the Canadian boundary, they can be enumerated as follows: Mount Shuksan, Mount Baker, Mount Logan, Glacier Peak, Mount Rainier, Mount Adams, and Mount St. Helens (all in Washington); then across the Columbia River in Oregon are Mount Hood, Mount Jefferson, the Three Sisters, and so on down to Lassen Peak and Mount Shasta in California. Crater Lake occupies the great crater-like depression that was formed by the destruction of another large volcano, the former Mount Mazama of prehistoric time.

All of these peaks, and many others, have seen their most active days. Some of the peaks still show slight volcanic activity. Lassen Peak in recent years has been the most active of all. Even Mount Rainier occasionally gives forth steam and hot gases from its summit, but not enough to melt the snow fields which enshroud its crest. Most of the peaks just mentioned support glaciers which radiate like rivers of ice down the mountain flanks.

The three sketches above are typical of the many Cascade volcanoes. They are all simple peaks rising above a vast upland. Mount Rainier attains an elevation of over 14,000 feet. Most of the others rise to 10,000 feet or so. This is 3,000 to 4,000 feet above the Cascade upland.

To the geologist studying topographic maps, the presence of a well defined radial drainage pattern around a mountain peak is evidence that the peak is a volcano even though it may be covered with snow and may exhibit no distinct crater. Certain other famous peaks, like Pikes Peak and Mount Washington, do not exhibit radial drainage because they do not stand isolated as volcanoes do but are parts of a mountain range. Mount Washington, for example, is one of the peaks of the Presidential Range.

EXAMPLE 46

The Problem

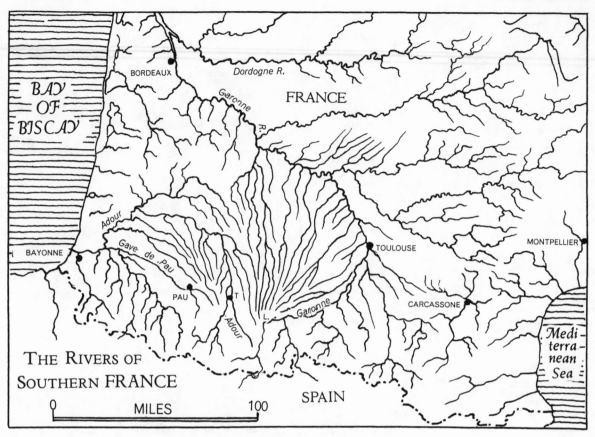

RIVERS. *River Patterns. More Radial Rivers. Southern France.*

In southern France just north of the Pyrenees is a region drained by the tributaries of two rivers, the Garonne and the Adour. The many tributaries of these streams radiate outward from a point at the northern base of the Pyrenees near the cities of Tarbes and Lannemezan. The position of these cities is indicated by the letters "T" and "L" on the above map. The radial pattern of these streams contrasts in a noteworthy manner with the irregular drainage pattern of the streams on the rest of the map.

In the preceding example we examined the radial drainage pattern of rivers on a large volcano, Mount Rainier. In that instance we observed that the streams radiated outward like the spokes of a wheel throughout the whole circumference of the circle. In the instance depicted on the above map the radial pattern involves only half the circular area. The remarkable characteristic about the stream systems in southern France is the fact that even the smallest tributaries form part of the pattern. There are no branches running off in other directions.

Another observation may be made concerning the size of the areas involved. According to the scale of the map above, the circular area in which the radial drainage occurs is more than 100 miles in diameter. In the preceding example the diameter of the area involved was less than ¼ of that amount. Nor did all the smaller streams in the preceding case conform so closely with the radial pattern of the whole region. It seems safe to say that the radial pattern of the two regions is due to two quite different causes.

In many other parts of the world radial drainage patterns similar to that in southern France also occur. They are, however, most of them, too small to show clearly on small-scale atlas maps. Fairly good examples may be detected on maps of India where the lesser tributaries of the Ganges and the Indus rivers drain away from the lower slopes of the Himalayas. Southern California has some splendid examples, clearly shown on the large-scale topographic maps, in the basin-like areas amongst the Coast Ranges, not far from Los Angeles. In Peru, the tributaries of the Amazon, notably the headwaters of the Madeira, the Purus, and the Juruá, flowing away from the eastern slope of the Andes, exhibit a fairly good radial pattern on a large scale.

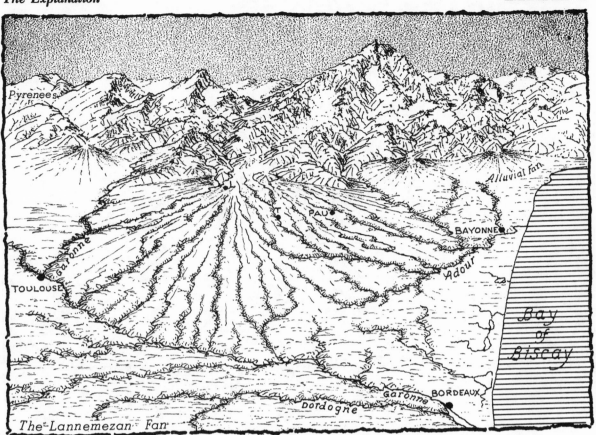

The above sketch is a view looking south toward the Pyrenees, from above the region near Bordeaux. It is turned therefore in exactly the opposite direction from that of the map on the preceding page. This has been done in order to provide a more direct view toward the northern base of the Pyrenees and to exhibit more clearly the fan-like slopes which emanate from the mouths of some of the valleys where they emerge from the mountains. The largest fan of all is the one drained by the tributaries of the Garonne and the Adour river systems. Having its apex near the town of Lannemezan, it is known as the "Lannemezan Fan."

Fan-shaped slopes of the type illustrated above are known as "alluvial fans." Alluvial fans form where streams heavily loaded with sediment come out from the mountains onto the plains. The sediment is there deposited in the form of a fan. A series of fans like those at the base of the Pyrenees may coalesce to form a long sloping plain. The High Plains portion of the Great Plains of Colorado and Nebraska represent a vast alluvial plain built out from the eastern base of the Rocky Mountains. Similar extensive fan-like slopes occur in northern Italy south of the Alps, and also in northern India south of the Himalayas. In Nevada and in Utah the mountain ranges of the Great Basin are rimmed by alluvial fans. Salt Lake City is located on a fan of that type.

The alluvial fans of southern France, like many others throughout the world, were formed during Glacial Time. In those days the glaciers of the Pyrenees were much larger than they now are. Plowing through their valleys, they produced a vast quantity of pulverized rock which the streams carried down as far as the plains. There, because of the more level ground, the speed of the flowing water was slowed down. As a result of this the streams had to deposit their loads of sediment. At the present time the glaciers are relatively inactive. The streams no longer bring down any material to be deposited. In fact, it is just the reverse. The streams are now eroding. They are forming sharply cut valleys in the long alluvial slopes. The valleys indeed are so steep-sided that they can be crossed only with great difficulty. The easiest way to go from one part of the alluvial fan to another is to go to the head of the fan at Lannemezan and then go out again in a new direction.

EXAMPLE 47

The Problem

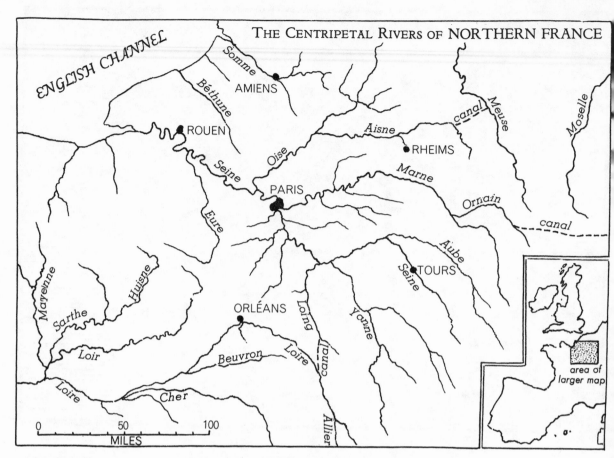

RIVERS. *River Patterns. Centripetal (Center-Flowing) Streams. Northern France.*

Almost any map of northern France shows that the tributaries of the Seine River System come from all directions and flow toward a common point which is about where the city of Paris is situated. Instead of radiating outward, as in the case of Mount Rainier, described in Example No. 45, the streams of northern France do just the opposite. The French streams exhibit a roughly radiating pattern like the spokes of a wheel. They flow, however, toward the center, and not away from it. Little wonder that Paris should hold so commanding a position. Like a spider sitting in the middle of its web, Paris is sensitive to all that goes on in the large drainage basin of the Seine River.

Perhaps the most important tributary to the Seine is the Marne River, which joins it right at the city of Paris. Indeed, the Marne is even more important than the above map alone would indicate, for its headwaters have been connected with the Rhine River by the Rhine-Marne Canal. This brings the whole of Alsace-Lorraine into close relationship with the Paris area.

Another very important tributary is the Oise-Aisne System, which reaches out into northernmost France. Here again another canal connects the Aisne with the Meuse River and thus with the important coal fields of northern France and Belgium.

Note next the upper Loire River. The Loire is not tributary to the Seine. It serves, however, much the same purpose. The upper Loire and its tributary, the Allier, flow north from the central upland of France directly toward Paris before turning west near Orléans. But note that the Loire is connected with the Loing by a canal, thus bringing the heart of central France into direct communication with the Paris area.

In few other regions of the world is the combination of river systems such as to make it almost inevitable that a great city and national capital should be established in so definite a location.

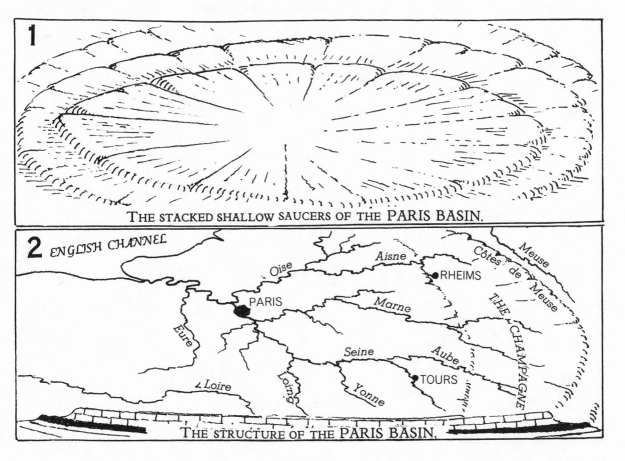

1

THE STACKED SHALLOW SAUCERS OF THE PARIS BASIN.

2 ENGLISH CHANNEL

Oise — Aisne — Côtes de Meuse — Meuse — RHEIMS — Marne — THE CHAMPAGNE — Eure — PARIS — Seine — Aube — Loire — Loing — Yonne — TOURS

THE STRUCTURE OF THE PARIS BASIN.

Northern France is shaped like a saucer. In fact, it resembles several flat saucers or plates stacked one on top of another. Were we to count them all we should find at least eight of them. The largest, the one at the bottom, reaches as far east as Germany. The smallest one, the one on the top, is in the center and constitutes most of the area with which we are here concerned.

Because of its form, this region is known geologically as the Paris Basin. The basin-like shape explains why most of the streams flow toward a common center. In addition to the numerous streams which flow down the inside of the basin, there are certain other streams which occupy the lowland areas between the edges of the several saucers. The Meuse is a good example of this. Its position is shown on Map No. 2 above. The scarp overlooking the Meuse is known as the Côtes de Meuse. The French called these scarps "coasts" because they were thought to have been formed by the sea.

Farther east a stream, similar to the Meuse, is the Moselle, which flows at the base of a scarp known as the "Côtes de Moselle." One of the broad lowland areas between two of the saucers is known as the Champagne, or the Plain. Still another lowland area farther east is called the Woëvre. The Forest of Argonne, famous as one of the battlefields in the First World War, is the name given to one of the upland areas that forms the rim of one of the saucers.

The structure of the Paris Basin is not entirely a simple one. Nevertheless, in spite of its long geological history and the complex nature of the adjustment of the streams to the rock structure, the centripetal pattern of the main drainage lines still persists, thus suggesting on even ordinary maps the basin-like arrangement of the geological formations. See also Example 9 for another geological basin.

EXAMPLE 48   *The Problem*

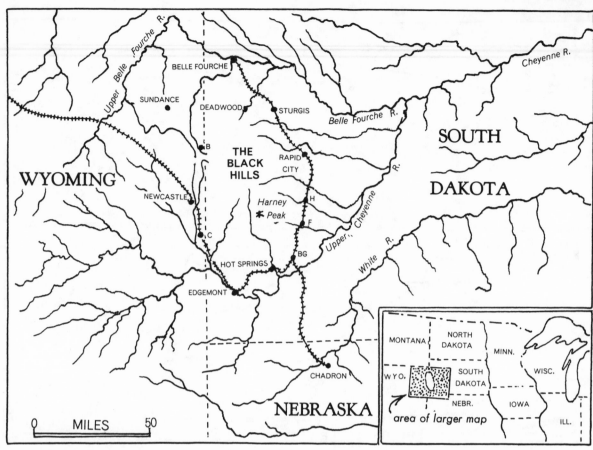

RIVERS. *River Patterns. Ring-Like or Annular Rivers. South Dakota.*

In western South Dakota, and surrounding the Black Hills, is a river system whose tributaries form almost a complete circle. This river plan can be detected on almost any large map of the United States. The automobile road maps do not show it very well because some of the streams have been omitted. The above sketch map depicts the stream pattern as it appears on one of the maps of the National Geographic Society. Some of the railroads have been indicated also, as well as several of the towns. A railroad runs around the eastern base of the Black Hills, and occupies a continuous valley in which are located half a dozen or more small cities. This valley is drained by a number of small wet-weather tributaries. The corresponding valley on the west side of the Black Hills is drained by two streams—one flowing to the north toward the town of Belle Fourche, and the other flowing to the south close to the town of Newcastle.

All of the various small streams which make up this ring-like pattern belong to the Cheyenne River drainage system. The Cheyenne River flows to the east to join the Missouri. The Upper Cheyenne wraps itself around the southern half of the Black Hills. In almost exactly the same manner the Belle Fourche, its northern tributary, wraps itself around the northern half of the mountain area.

In addition to the relatively small circular pattern just described, there is a second and outer series of streams which together form a more or less complete circle. Three streams in particular are involved; namely, the White River, the Upper Cheyenne, and the Upper Belle Fourche. Further study of the map will reveal still more evidence of annular drainage lines, particularly in the smaller tributaries.

The letters designating the towns are used again on the two explanatory maps on the next page, in order to assist in correlating the several maps.

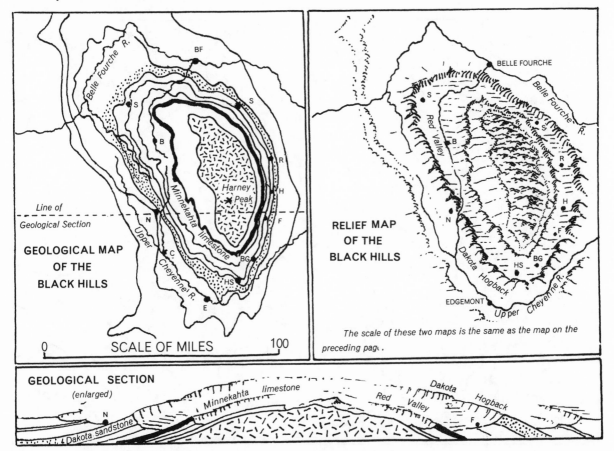

GEOLOGICAL MAP OF THE BLACK HILLS

Line of Geological Section

RELIEF MAP OF THE BLACK HILLS

The scale of these two maps is the same as the map on the preceding page.

SCALE OF MILES

GEOLOGICAL SECTION (enlarged)

The annular or ring-like arrangement of streams in the Black Hills region is explained by the two maps above. The region is a dome-like uplift, or blister-like swelling of the earth's crust. It is actually more than that, however. It is a blister which has been eroded. It has been worn down to reveal the underlying layers of rock. The effect is much like that obtained by cutting off the end of an onion so that the inside layers can be seen. The blister has not been cut off to a level plain. This is because the different underground layers of rock are not all alike. Some are more resistant to erosion than others. The resistant layers produce ridges which form rings around the dome. The weaker beds form lowlands which form similar rings, parallel to and alternating with the ridges. These rings of ridges and lowlands are indicated on the Relief Map above, the one on the right. Some of the rivers have been added to this map in order to show how they correspond with the lowland belts. The railroads and towns occupy the lowlands also.

The Geological Map above, at the left, shows the ring-like pattern of the geological outcrops. The two formations which produce the two most conspicuous ridges are emphasized. One is shown by dots; the other is shown by a heavy black line. Each one of these formations produces a ridge which is known as a "hogback." The outer one of the two hogbacks is known as the "Dakota Hogback." It is formed by the Dakota sandstone. Just within this circular ridge is the largest of the circular lowlands. Underlying this lowland are some weak red rocks. The soil in this lowland is therefore red, and the lowland is called the "Red Valley." Because of its ring-like form it is known also as the "Racetrack," a rather large racetrack, to be sure, for it is more than 100 miles from end to end.

Inside the racetrack is another hogback. This is produced by the Minnekahta Limestone, the formation shown by the heavy black ring on the Geological Map. The hogback produced by this formation is not so high as the Dakota Hogback, which rises 500 feet above the Red Valley.

EXAMPLE 49   *The Problem*

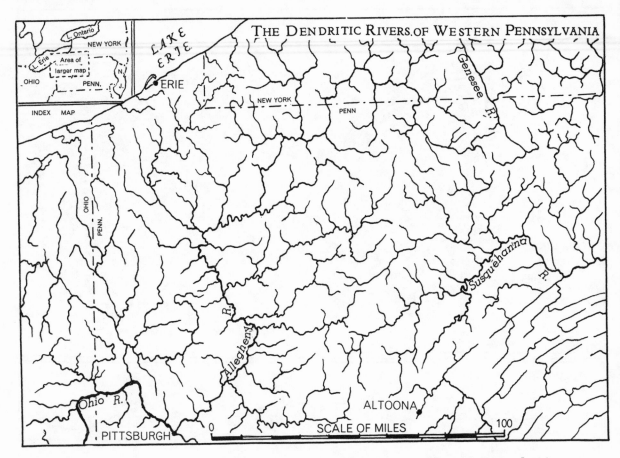

RIVERS. *River Patterns. Irregularly Branching or Dendritic Streams. Western Pennsylvania.*

The rivers of northwestern Pennsylvania are shown on the above map as they appear on many atlas maps as well as on such maps as the National Geographic Society map of the United States. The culture and most names have been omitted in order that the drainage lines may be more easily recognized. The dendritic or irregularly branching type of drainage is well shown here. The streams run in all conceivable directions. This might be called a drainage pattern without any particular arrangement. Perhaps at first thought it might be supposed that this is the way all rivers flow. Such, however, as we have already observed, is not the case. For contrast, there is included in the southeastern corner of the map a small area exhibiting trellis drainage pattern, the type of drainage common in the Folded Appalachians, and already described in Example No. 43.

Several river systems participate in the dissection of this northwestern corner of Pennsylvania. There are the headwaters of the Susquehanna, the Allegheny and other tributaries of the Ohio, and the Genesee. Each of these has many branches, and there are vast numbers of smaller streams and brooks that are not shown on maps of this scale. They may be seen on the large-scale topographic maps which are about 30 times as large as the map presented above. There are untold thousands of these smaller tributaries likewise flowing in all possible directions, so that no pronounced pattern is detectable.

As we puzzle over maps showing dendritic drainage patterns, and consider again some of the other kinds of drainage patterns that have already been described, we are forced to conclude that this tendency of streams to flow in all directions is to be explained by the simple statement that there is nothing to cause them to flow in any particular way. Their locations are determined pretty much by chance. It remains, however, for us to determine just what kind of rock structure will bring about this effect.

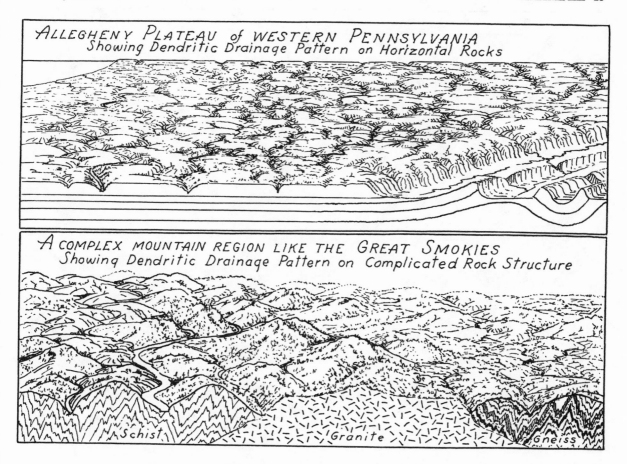

ALLEGHENY PLATEAU of WESTERN PENNSYLVANIA
Showing Dendritic Drainage Pattern on Horizontal Rocks

A COMPLEX MOUNTAIN REGION LIKE THE GREAT SMOKIES
Showing Dendritic Drainage Pattern on Complicated Rock Structure

Schist   Granite   Gneiss

The above sketches illustrate the two most common geological conditions which cause rivers to have dendritic drainage patterns.

Western Pennsylvania is a plateau. The uppermost sketch above shows somewhat diagrammatically what this plateau is like. The streams have cut deep valleys below the plateau surface, and hence most of the actual streams are not visible. It is clear, however, that the region is made up of numerous hills. The valleys which separate the hills from each other branch off in all directions. In the foreground is a geological cross section. This shows that the formations which underlie the plateau consist of horizontal layers of rock. Because the beds are horizontal there is only one kind of rock which appears at the surface of the ground. This means that it is as easy for streams to flow in one direction as in another.

At the right-hand end of the cross section is a portion of the Folded Appalachians. In this part of the area the streams have a trellis pattern because some of them flow along the belts of weaker and more easily eroded beds of rock. The more resistant beds form the ridges between the valleys.

In the lower sketch is presented a quite different kind of region. This region also has a dendritic drainage pattern. The country is more rugged than the Allegheny Plateau. It is similar to such regions as the Great Smoky Mountains of the Southern Appalachians. Other similar mountains are the Salmon River Mountains of Idaho. The rocks in this second region are not horizontal. They consist instead of minutely crumpled formations such as gneiss and schist, or of great masses of crystalline rock like granite. These rocks, by and large, have all about the same degree of resistance. The result is that there is little control upon stream position. The numerous crenulations in the rock, moreover, cause the streams to flow in just about all directions.

EXAMPLE 50   *The Problem*

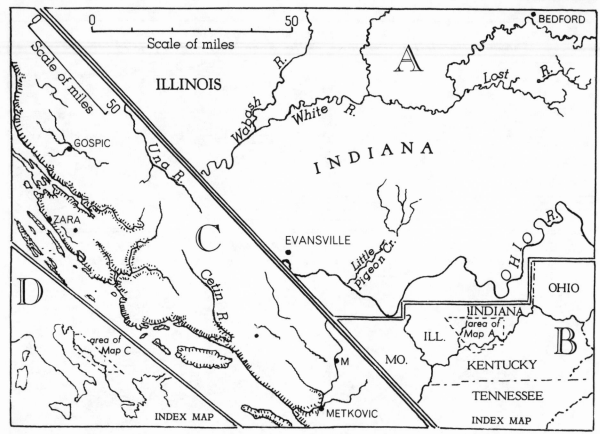

RIVERS. *Interrupted or "Lost" Rivers. Southern Indiana; Yugoslavia.*

As we examine maps of various sorts we occasionally come across some parts of the world where the streams appear to be interrupted in their courses. They stop without joining or flowing into any other streams. This condition may be the result of any one of several causes. One type of such stream is represented by the two examples shown above, taken from widely separated areas.

Streams of the type represented above are relatively short, usually only a few miles in length; and this is far too small to be shown on ordinary maps. At the outset, then, it can be said that in our perusal of maps, except for the very large-scale topographic sheets, we are not apt to run across interrupted streams. Nevertheless, streams of this type are not at all unusual, and it is for this reason that these examples are included here. Inasmuch as it is agreed that in this book topographic maps are assumed not to be available to the reader, reference will therefore be made to other more accessible material.

Take first the "lost" rivers of southern Indiana, Kentucky and Tennessee. Fortunately two or three of these are shown on the National Geographic Society map of the Southeastern United States, approximately in the form indicated on the above sketch map, Figure A. The region where interrupted streams are common is shown on the Index Map, Figure B. Notice, on Figure A, the small stream south of Bedford, Indiana. It is entirely detached from any other stream system, and appears to be flowing into a small lake. Two tributaries join to make this stream, which is, all told, about 15 miles long. Another "lost" stream of similar length may be noted east of Evansville, where the headwaters of Little Pigeon Creek fail to join the main stream. It is not at all uncommon for streams of this type to be named "Lost River."

A second example has to do with the streams in the Dalmatian area of Yugoslavia, as shown on another one of the National Geographic Society maps. Most maps of Europe as a whole fail to show these small broken streams. Figure C above represents a few of these streams, several of which are as much as 25 miles long. There are literally hundreds of them scattered throughout Dalmatia.

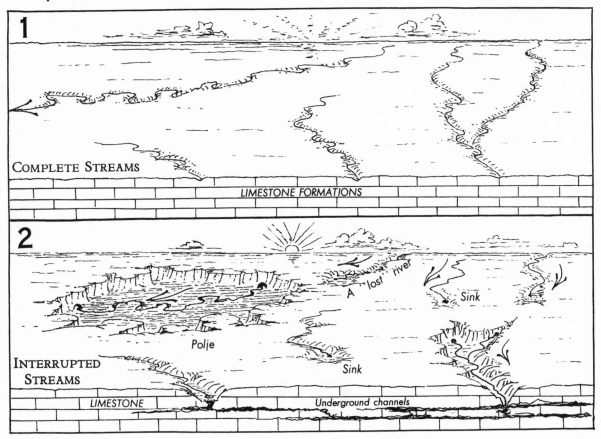

Interrupted or lost rivers frequently occur in limestone regions. There are other types of interrupted or broken rivers also, but the examples cited on the preceding page are in the limestone category.

Limestone is a relatively soluble rock compared with other formations like shale and sandstone, or igneous rocks like granite. Water falling upon limestone areas tends to seep down through cracks or joints and gradually to dissolve the limestone to form openings and channel ways. These openings through which the water goes underground are called "sinks" or sinkholes. Drainage from the surface of the country, even though at first there may have been a unified stream system (as in the upper figure above), gradually becomes disintegrated (as in the lower figure). Streams which at one time were parts of a regular system seep down through the sinks into underground passages, possibly flowing for many miles before regaining their former main stream farther down the valley. Or perhaps such a detached and underground stream may even reach the sea independently. At any rate it is no easy matter to know what becomes of a stream once it has disappeared underground. In some instances these streams can be traced by means of dyes or even by throwing sawdust into them.

The island of Barbados, which is almost entirely a plateau of coral limestone, has virtually no surface streams whatever, its only river being in a non-limestone area. Nevertheless Barbados has abundant reservoirs, or really rivers, of underground water. The surface of the island contains many depressions which carry off the rainfall before streams have time to form.

In the so-called "Karst" region of Dalmatia, the streams occur on the floors of large depressions, called "poljes." These are several miles in length, and are rimmed on all sides by wall-like valley sides. The streams emerge on one side of the depression from some underground channel and, after traversing the valley floor, disappear again into another cave-like opening, thus becoming a "lost" stream. This is illustrated in Figure 2 above.

EXAMPLE 51

The Problem

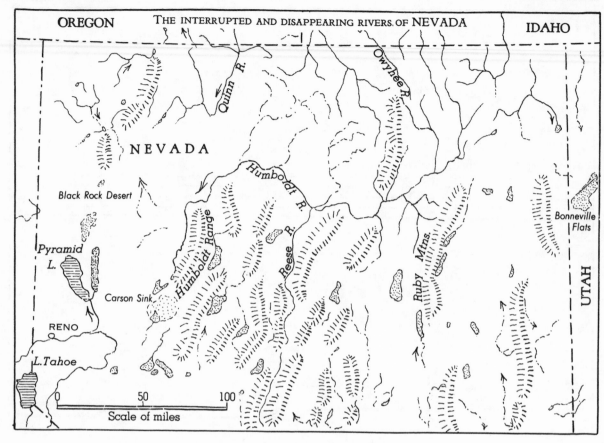

THE INTERRUPTED AND DISAPPEARING RIVERS OF NEVADA

RIVERS. *More Interrupted Rivers. Nevada; Mexico; Persia; Australia.*

Another type of interrupted river, quite different from those found in limestone regions, occurs in Nevada and the adjacent states of Utah, California, and Oregon. From the appearance of the map alone, one would never suspect that many of these streams are dry most of the time. They are usually mere beds of boulders with here and there a water hole. Many of these rivers end in lakes or so-called "sinks," or "playas." Such lakes are very shallow, and vary tremendously in area, in accordance with the flow of the rivers entering them. These lakes are apt to be briny too or, when dry, merely salt flats. Such is the famous Bonneville Salt Flat where the automobile racetrack is situated.

The longest of these Nevada rivers, which come to an end without reaching the sea, is the Humboldt. Because this stream usually contains a fair amount of water, the valley of the Humboldt served as the "emigrant trail" to the West during the Gold Rush days. Along this valley now run the Union Pacific Railroad and U.S. 40, the Lincoln Highway.

A perusal of the above map reveals numerous shorter streams which follow very undecided courses, petering out from time to time in the middle of nowhere.

Some very interesting interrupted rivers may be observed also on almost any map of Australia. Like those of Nevada, they are dry most of the time, but once or twice each year their channels carry a mighty flood. Such is the Todd River which flows through Alice Springs. Except for a pool of water here and there, it can be crossed dryshod most of the year. Although dry, the river courses of central Australia are marked by continuous belts of eucalyptus trees, whose roots penetrate to the reservoir of water in the gravel beneath the dry river bed. In this manner also, the aborigines are able to survive in a country which appears waterless. Cattle, however, during the dry periods, succumb by the thousands. This is due not so much to thirst, for the cattle stations are all provided with wells and windmills, but to the lack of forage in times of drought.

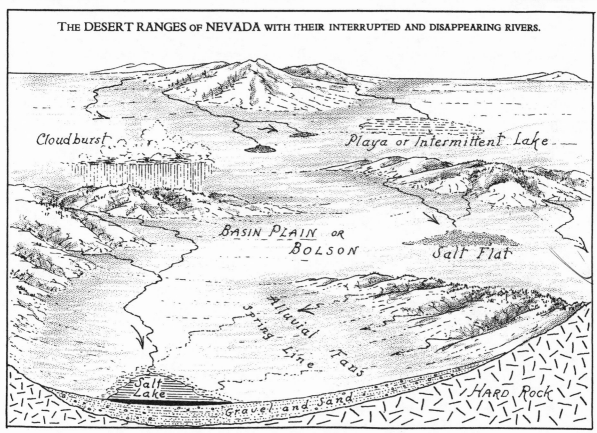

THE DESERT RANGES OF NEVADA WITH THEIR INTERRUPTED AND DISAPPEARING RIVERS.

The rivers of Nevada, Persia, Australia, Central Asia, and similar areas, all occur in regions of deficient rainfall where the precipitation is around 5 inches or less a year, and rarely as much as 10 inches. Most of this rainfall comes in short spells of the thunderstorm type, and much of it is concentrated upon the usually small and scattered mountain ranges. This results in brief periods of flash floods and long periods of drought.

Under such conditions, because the streams are not through-flowing, they tend to build up extensive alluvial deposits or "fans" which spread from the foot of the mountains out onto the basin floors. Many streams emerge from the mountains, where the rainfall is more generous than elsewhere, and flow out upon the alluvial sand and gravel deposits. There they lose themselves in the porous material, perhaps to reappear farther down the slope at the so-called "spring line." This presupposes that there is enough water underground to make this possible. Otherwise the thirsty gravel soaks up the river and it is lost for good.

Some streams, of course, survive until they reach the lowest part of the intermontane basins. Here lakes result, and if there is sufficient underground water at this level the lake may be more or less permanent. Even so, because of the intense evaporation, and the varying amounts of water brought into the lake, it changes greatly in volume and therefore in area. Such lakes, too, in the course of time, become salty, not only with the concentration of sodium chloride but with many other chemical compounds as well, such as gypsum and potassium salts. If there happens to be a chain of these lakes, the last and lowest member of the chain will be the saltiest one. One of the most famous lakes of this kind is Great Salt Lake. Most of the Nevada lakes shown on the opposite map are even more saline in composition. Once in a while a lake like Lake Eyre in southern Australia, which is usually just a vast salt flat, increases sufficiently in volume to overflow and eventually to reach the sea, but this happens only about once in a lifetime.

EXAMPLE 52

The Problem

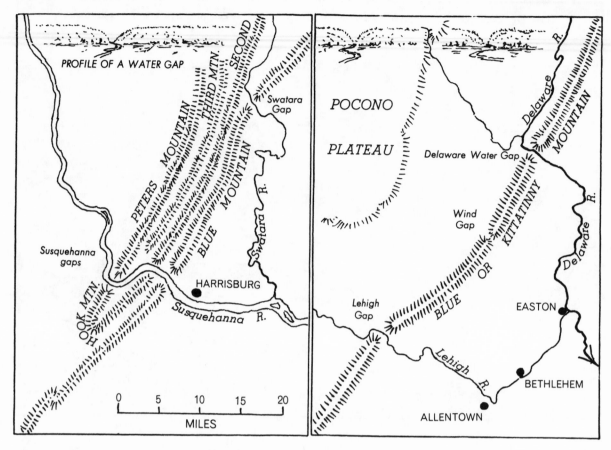

RIVERS. *Water Gaps. Delaware Water Gap; Susquehanna Water Gaps.*

"The mighty river, tumbling through the mountains, finally mustered all its strength and burst through the great range which had been blocking its way." Thus the famous Delaware Water Gap is sometimes described. To believe that rivers, like a human adventurer, suddenly find themselves confronted with a barrier through which they have to break a passage is fanciful indeed. Water gaps do not come about in that way. Let us look at a couple of maps of some well known water gaps in the eastern United States and become more familiar with their characteristics before we try to decide how they were formed. We know that the rivers themselves formed the gaps, but just how did they go about it?

Each of the above maps shows one or more long, almost unbroken ridges, transected here and there by rivers, both large and small. These ridges are indeed veritable mountains, rising a thousand feet or more above the low ground on either side. If seen from a distance they would appear flat-topped, as shown by the profiles above the maps. Besides the water gaps which interrupt the continuity of the ridges, there are to be seen here and there low sags or so-called "wind gaps" not utilized by any stream.

The first map above shows the Susquehanna River cutting across three main ridges. This seems a bit remarkable because, by swinging a few miles to the southwest around Hook Mountain, it could avoid two of the ridges entirely. The Swatara River behaves illogically also because, after flowing along an open valley for several miles, it turns abruptly through its gap across Blue Mountain, although its original valley continues far to the southwest. The Delaware River on the next map behaves in precisely the same way. Why do these rivers abandon perfectly good valleys and cut through high mountain barriers? Obviously some far-reaching explanation is called for.

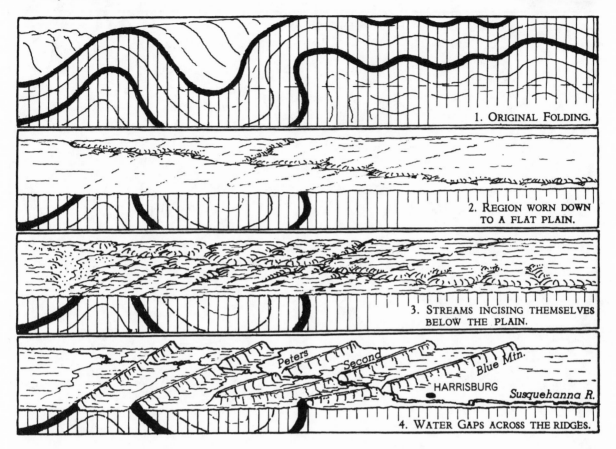

1. ORIGINAL FOLDING.

2. REGION WORN DOWN TO A FLAT PLAIN.

3. STREAMS INCISING THEMSELVES BELOW THE PLAIN.

4. WATER GAPS ACROSS THE RIDGES.

The above sequence of diagrams shows how water gaps are formed across a succession of long, more or less parallel, even-topped ridges.

In the first diagram, marked "original folding," we see a series of geological formations that have been folded into great arches and troughs (anticlines and synclines). This represents in a much simplified way a typical part of the Appalachians. The formations include resistant beds of sandstone and conglomerate alternating with weaker beds of limestone and shale. Each formation may be hundreds of feet in thickness, and the folds may have reached thousands of feet in height.

In the second diagram we see the region some geological ages later. It has been worn down practically to sea level to form a vast plain sloping gently eastward to the sea. Across the plain rivers are flowing. On the surface of the plain the "outcrops" of the truncated underlying folded formations are barely perceptible, being covered by alluvium and the soil produced by the accumulated weathering of the ages.

In the third diagram we see the same region a little later. The streams are starting now to incise themselves below the surface of the land, doubtless because there has been a slight elevation of the region above the sea with resulting "rejuvenation" of the drainage system. The streams form continuous valleys of uniform depth. There are as yet no ridges to cut through. However, we see that some of the tributaries are discovering the belts of weaker rock. As rapidly as the main streams incise themselves the tributaries readily keep pace with them in this weaker material.

In the fourth diagram we note that the tributaries have succeeded in forming broad valley belts where the weaker rocks outcrop. The resistant beds remain as ridges. The gaps across them have been cut from above downward. Rivers working in this way are called "superposed." The ridges themselves have resulted from the removal of the weaker rocks on either side. They have not been pushed up, as one might at first think.

EXAMPLE 53

The Problem

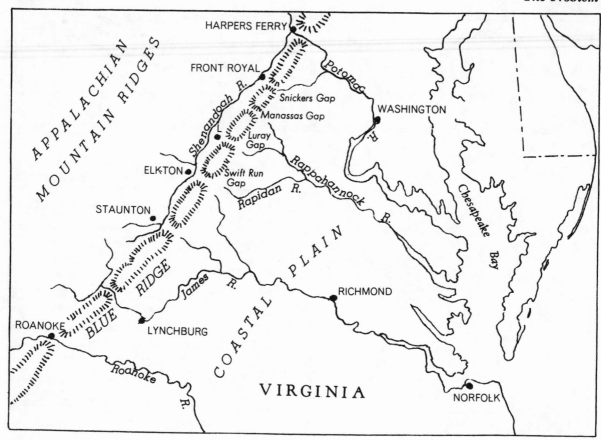

RIVERS. *Water Gaps and Wind Gaps. The Blue Ridge, Virginia.*

From one of the road maps published by the state of Virginia the above simple sketch map was prepared. It emphasizes only a few of the significant features. The feature deserving especial consideration is the Blue Ridge with its several water gaps and wind gaps. The water gaps are those formed by the Potomac, the James, and the Roanoke rivers. The wind gaps all occur between the Potomac gap at Harpers Ferry and the James River gap west of Lynchburg. All of the various gaps are used by highways of one sort or another which cross the Blue Ridge. Several of them are used by railroads as well, particularly the water gaps because of their lower elevation. The main line of the Baltimore and Ohio Railroad goes through the Harpers Ferry gap of the Potomac River; and the main line of the Chesapeake and Ohio utilizes the James River gap west of Lynchburg. The wind gaps in general stand at elevations above those of the water gaps. However, the wind gaps are all well below the level of the crest of the Blue Ridge, by as much as 2,000 feet or so.

Several of the wind gaps are labeled on the above map, as follows: Snickers Gap, Manassas Gap, Luray Gap, and Swift Run Gap.

The Potomac, the James, and the Roanoke all have their headwaters west of the Blue Ridge, whereas many of their smaller tributaries, such as the Rappahannock and the Rapidan, rise along its eastern slopes. A rather noteworthy tributary of the Potomac River is the Shenandoah. It flows along the western base of the Blue Ridge, and has its headwaters close to the James River.

West of the Blue Ridge are many other linear and parallel ridges of the Appalachian System. These other ridges, like the Blue Ridge, have many water gaps and wind gaps whose origin is doubtless the same as that of the Blue Ridge gaps. The Blue Ridge region, therefore, may be taken as a type example to explain the many wind gaps and their relation to the water gaps of the entire Appalachian System.

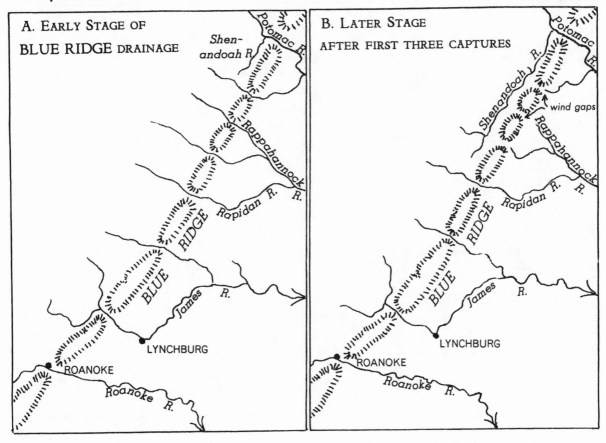

We have already seen in the preceding example how some of the famous water gaps of the world have come about. The problem in this example is to explain the so-called "wind gaps" through which no stream now flows.

On the above map, Figure A shows an early stage in the drainage of the Blue Ridge area. There were at that time quite a number of streams which crossed the Blue Ridge in water gaps. There were not only the Potomac, the James, and the Roanoke, but also such smaller streams as the Rappahannock and the Rapidan. The Shenandoah was at that time only a small tributary of the Potomac. The Potomac was the master stream of the entire region. Its large volume resulted from its extensive headwater drainage area back in the Appalachians. This enabled it to cut down its valley across the Blue Ridge to a depth lower than that of any of the other transverse streams. This in turn enabled the Shenandoah to cut lower and to work headward toward the southwest along the limestone belt of what is now the Shenandoah Valley. As it worked headward it impinged upon the upper courses of the Rappahannock and other rivers which, because of their smaller volume, could not cut down their gaps to so great a depth as that of the Potomac. Their headward portions were therefore "captured" by the Shenandoah and diverted into its drainage system.

Figure B shows one of the later stages after three of the earliest captures have occurred. From this it is easy to see how the Shenandoah continued on its career of "piracy" and captured the Rapidan and the smaller tributaries of the James River. As time goes on, it is safe to predict, the Shenandoah will next take over the main headwaters of the James and still later those of the Roanoke. This process explains the relatively small number of large transverse streams across the Appalachians at the present time, as well as the numerous wind gaps that occur along the crests of the Appalachian ridges.

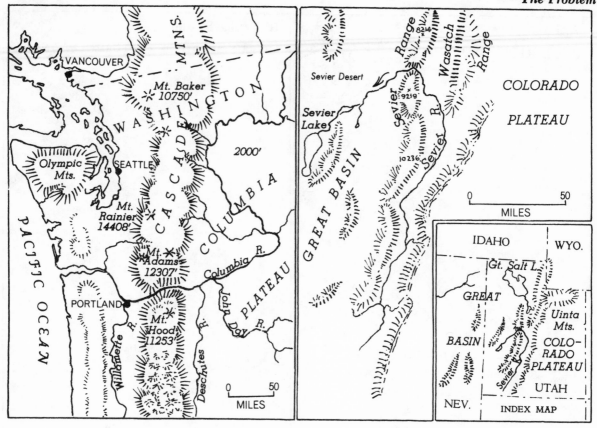

RIVERS. *More Water Gaps. Columbia River Dalles Across the Cascade Mountains; Sevier River Gap, Utah.*

The Columbia River, after draining the large basin-like area of central Washington known as the Columbia Plateau, turns west and cuts through a deep gorge across the Cascade Mountains. The Cascades, as shown above, form a barrier, several thousands of feet high, which rims the western side of the vast Columbia River Basin in Washington and Oregon. This basin, best known as the Columbia Plateau, embraces an area all told, in Washington, Idaho, and Oregon, some five times that of New York State. The elevation of this plateau is greatest in its eastern part, adjacent to the Rocky Mountains. The plateau level descends to about 2,000 feet or so in central Washington. The Cascade upland rises abruptly above the Columbia Plateau to an elevation of 10,000 feet or more, and is surmounted by several graceful volcanic cones going up several thousand feet higher. Here the Columbia River gorge transects this mighty range for a distance of 50 miles. The eastern end of the gorge is known as The Dalles. It is in this gorge that the great Bonneville Dam has been built to block the Columbia River, thus accomplishing what the Cascades themselves were apparently unable to do. Farther south, we may note on appropriate maps that the Cascade uplift is cut through by two other streams, the Klamath River and the Pitt River in that portion which is called the Klamath Mountains.

Another unique gap is that of the Sevier River, which transects the Sevier Range in Utah, south of Great Salt Lake. The Sevier Range is one of the numerous relatively short block mountains which run like caterpillars across the Great Basin. Although a few of these ranges have high gaps and passes near their crests, almost none of them is crossed by rivers. For one thing, rivers are very scarce in the Great Basin, and most of them are dry much of the time anyway. Besides that, the ranges are so discontinuous that the rivers seem to have no trouble flowing around them. Why the Sevier River should choose to cut across the Sevier Range truly demands some explanation.

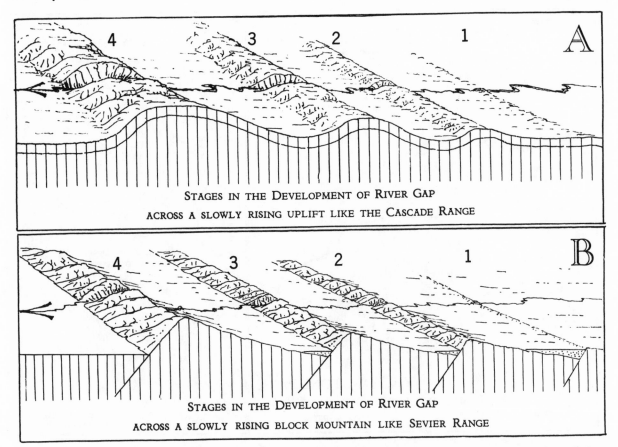

STAGES IN THE DEVELOPMENT OF RIVER GAP
ACROSS A SLOWLY RISING UPLIFT LIKE THE CASCADE RANGE

STAGES IN THE DEVELOPMENT OF RIVER GAP
ACROSS A SLOWLY RISING BLOCK MOUNTAIN LIKE SEVIER RANGE

The two water gaps presented in this example were formed in a manner quite different from that of the water gaps across the Appalachian ridges described on the preceding pages. The essential difference between the Appalachian ridges and such ranges as the Cascades and the Sevier Range is this:

The Appalachian ridges were formed because the weaker formations on each side of the ridge were worn away. The ridges were not pushed up individually.

The Cascades and the Sevier Range, on the other hand, actually represent uplifted masses or blocks of the earth's crust. They are not the result of erosion. In other words, the Columbia Plateau is not lower than the Cascades because it has been worn down that much.

We find upon extensive study of these regions and other similar areas in the world that the Cascade Mountains and the mountains of the Great Basin, such as the Sevier Range, have been slowly rising above the regions adjacent to them.

The Columbia River was there before the Cascades were ever formed. The upper diagram above shows the gradual growth of the Cascades across the path of the Columbia River. The uplift of the Cascades occurred so slowly that the Columbia was never blocked up, but maintained its course by cutting down its gorge as fast as the mountain mass was raised.

The Sevier River in Utah behaved in the same way. The Sevier Range differs from the Cascades in that it is a block mountain formed by faulting, and not an arch formed by folding such as the Cascades are. This contrast is shown in Figures A and B above. The Sevier River, it may be noted, is a more powerful and continuously flowing stream than most rivers of the Great Basin because it is adjacent to the Colorado Plateau, which induces a heavy rainfall in the headwaters of the Sevier River.

Streams like the Columbia and the Sevier rivers are known as "antecedent" streams because they preceded or antedated the mountains across which they flow. This is quite in contrast with the "superposed" character of the Delaware and the Susquehanna previously described.

EXAMPLE 55

The Problem

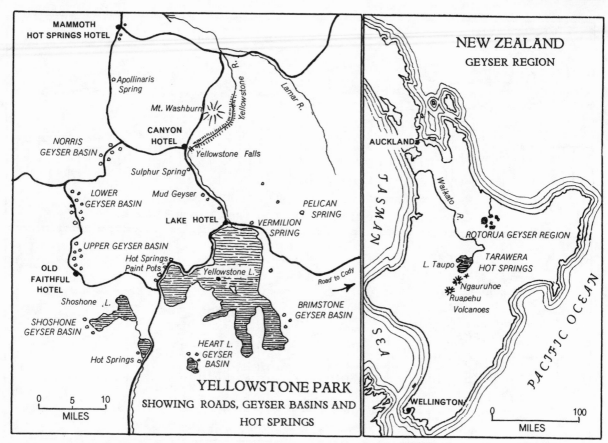

RIVERS. *River Sources. Hot Springs and Geysers. Yellowstone Park; North Island, New Zealand.*

The geysers and hot springs of the world, like springs of any kind, are parts of river systems.

Rivers derive their water from the rain which falls upon their drainage basins. A considerable amount of this rain does not flow into the rivers immediately but soaks into the ground to appear later along the stream courses in the form of seepages or springs. The exact location of seepages and springs depends upon the configuration of the ground and also upon the rock structure and other subsurface conditions. In volcanic regions, or in regions which have recently been volcanic, the water percolating into the soil may at depth encounter heated rocks. When this water emerges again it forms hot springs, or, if heated sufficiently to be turned into steam, it may take the form of geysers. Geysers, hot springs, mud springs, sulphur springs, boiling springs and pools are all manifestations of the same natural phenomena.

The problem before us in this example is not to explain the presence of geysers and hot springs in volcanic regions, but to go even further than that and explain if possible the distribution of these various springs within the volcanic area itself.

The Yellowstone Park geyser region and the New Zealand geyser region each embraces about the same number of square miles of country. Referring to the two maps above, we may note a number of similarities between the two regions. Each region still has one or more volcanoes: the ancient and eroded volcano of Mount Washburn in Yellowstone Park, and the still active volcano of Ngauruhoe with several neighboring ones in New Zealand. A large lake and several smaller lakes lie in each of these two regions, Lake Taupo of New Zealand being almost twice as large in area as Yellowstone Lake. The geysers of Yellowstone Park are distributed in several groups widely over the area, whereas the geysers of New Zealand are limited pretty much to the Rotorua region. Moreover, the total number of geysers in New Zealand is far less than that in the Yellowstone. Each region is in the nature of a broad lava plateau broken into several units and transected in a deep gorge or canyon by the stream which forms the outlet of the largest lake.

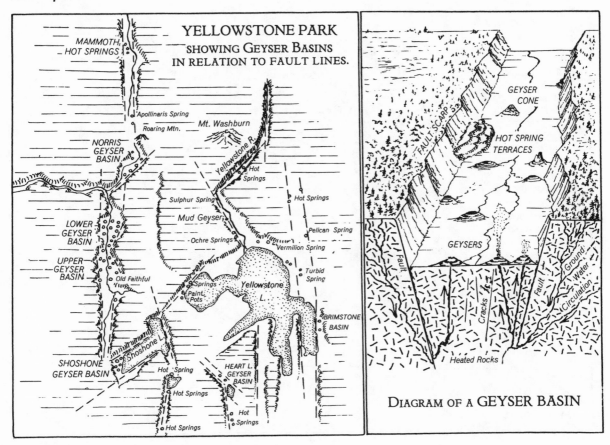

The geysers and hot springs of Yellowstone Park, as well as those of New Zealand and other parts of the world, occur along fault lines and grabens, or down-dropped trenches. The illustration above shows diagrammatically the main features of Yellowstone Park. We note that this region is a broad plateau broken into segments by fault scarps and trench-like basins. The main fault systems run north and south, and a less important group of faults runs in a northeast-southwest direction. The several arms of Yellowstone Lake fit nicely into this fault pattern. Shoshone Lake, likewise, occupies a couple of straight-sided grabens. The Norris Geyser Basin and the Upper and Lower Geyser Basins lie in similar grabens.

The geysers and hot springs are concentrated along these fault lines because the faults or cracks in the earth's crust provide channels through which the ground water can rise from the depths to which it has descended. Like all springs, geysers derive their water from the rainfall of the region. The small diagram at the right, above, illustrates schematically a small graben or geyser basin. The rain falling upon the plateau percolates downward through numerous small cracks and crevices in the rock until it encounters a fault plane along which it can again rise to the surface. The water rises because of the hydrostatic pressure behind it. Flowing springs thus result. The springs are hot if the water has encountered hot rocks during its journey. Geysers result if the water has been sufficiently heated to become steam. Many geysers erupt either spasmodically or more or less regularly, depending upon the nature of the underground passageways. Most geysers and hot springs build up cones of siliceous sinter, the silica being derived from the rocks with which the hot water has been in contact. Some springs, such as the Mammoth Hot Springs, deposit calcium carbonate material, derived from underground limestone formations. These deposits are often brightly colored, usually orange, red, or yellow, because of the algae which thrive in the hot water. If the spring ceases to flow, the algae die and the formations become white and chalky. Geysers sometimes become extinct too, leaving merely a dead sinter cone.

EXAMPLE 56

The Problem

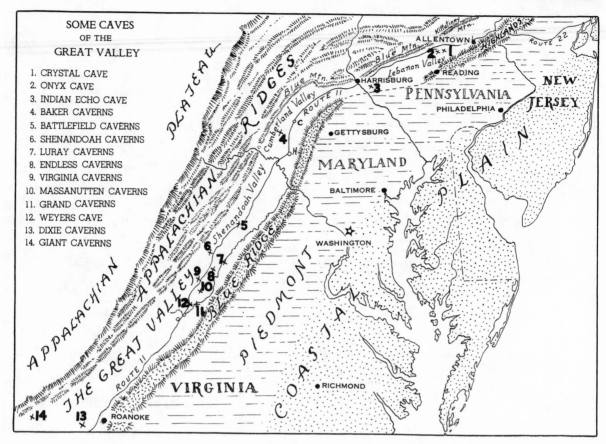

SOME CAVES
OF THE
GREAT VALLEY

1. CRYSTAL CAVE
2. ONYX CAVE
3. INDIAN ECHO CAVE
4. BAKER CAVERNS
5. BATTLEFIELD CAVERNS
6. SHENANDOAH CAVERNS
7. LURAY CAVERNS
8. ENDLESS CAVERNS
9. VIRGINIA CAVERNS
10. MASSANUTTEN CAVERNS
11. GRAND CAVERNS
12. WEYERS CAVE
13. DIXIE CAVERNS
14. GIANT CAVERNS

RIVERS. *Underground Rivers and Caves. The Cave Belt of Virginia and Pennsylvania.*

The above map is a compilation made from several road maps and similar available material. It shows the location of some of the commonly visited caves of the Eastern states. These caves all lie on one of the main inland routes between New York and the South: Route 22 between New York and Harrisburg and Route 11 which runs from Harrisburg south through western Virginia. The continuous valley through which these highways run is known in eastern Pennsylvania and adjacent New Jersey as the Kittatinny Valley. In Pennsylvania between Allentown and Harrisburg it is called the Lebanon Valley. Here we find three well known caves: Crystal Cave and Onyx Cave west of Allentown, and Indian Echo Cave not far from Harrisburg. Still farther south, in the region of Pennsylvania known as the Cumberland Valley, are the Baker Caverns. In Virginia, in the beautiful Shenandoah Valley section, are a score or more of caverns, of which a dozen or so are shown on the above map.

East of this great valley—and indeed it is actually known in its entirety as *the* Great Valley or Great Appalachian Valley, running from New York State to Georgia—is a high mountain belt. East of the Great Valley (the Shenandoah Valley) in Virginia is the Blue Ridge, which, extending northward, comes to an end near Gettysburg, Pennsylvania. In eastern Pennsylvania and in New Jersey the Great Valley is hemmed in on the east by a belt of Highlands which continue northward to form the Highlands of western New Jersey and farther on the Highlands of the Hudson. In the South the Great Valley is known as the Valley of East Tennessee. East of this area are the high mountains of the Great Smokies which continue northwards to form the Blue Ridge.

West of the "valley of the caves" or the Great Valley are the Appalachian ridges of Pennsylvania represented by Kittatinny Mountain and its southern continuation, Blue Mountain. The fact that all these caves lie in the Great Valley is worthy of note and therefore demands an explanation. All of this is an example of the systematic arrangement of topographic features so well displayed in the eastern United States.

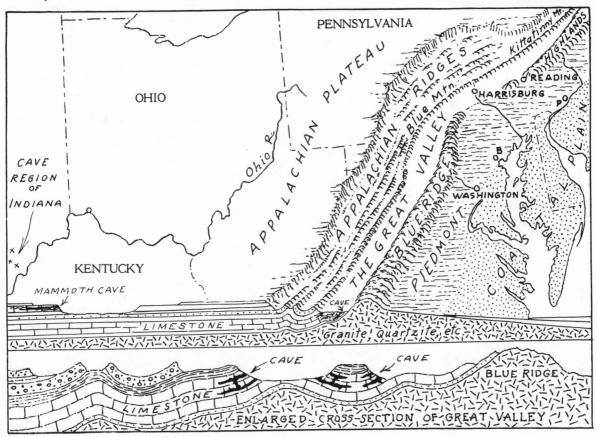

The purpose of this explanation is to give the reason for the long belt of caves that runs from Pennsylvania to Virginia. Incidentally, it will show also some of the differences between these caves and those which occur in the Mammoth Cave region of Kentucky. The above diagrammatic map, a kind which is frequently used to explain things in geology, shows the Great Valley cave belt in relation to the main features of the eastern states. It shows also along the front edge of the map a simplified geological cross section designed to represent the underground structure. At the very bottom is another section, somewhat enlarged, to illustrate better some of the details in the Great Valley where the caves occur.

The following features may be identified on the map as well as on the cross section: 1. the Blue Ridge, a high belt of resistant rocks; 2. the Piedmont and Coastal Plain, lying east of the Blue Ridge; 3. the Great Valley west of the Blue Ridge, a belt of folded limestones and weak rocks. It is in these limestones that the caves occur. 4. West of the Great Valley is the belt of Appalachian ridges, formed by resistant beds that overlie the limestones. 5. Still farther west are the Appalachian plateaus formed by the highest strata of all. Among these higher formations is the limestone in which Mammoth Cave occurs.

The Mammoth Cave limestone lies in horizontal layers, and Mammoth Cave therefore exhibits several levels that are reached, one to the other, by winding passageways. The caves of Virginia occur in folded, bent, or dipping layers. The cave rooms, therefore, are only occasionally horizontal. The warped and folded strata can be seen in the various passageways. Because the limestone "strikes" or outcrops in a northeast-southwest direction, the Virginia caves tend to be elongated in that direction also. On the other hand, the caves of the Mammoth Cave region sprawl out in all directions.

It is surprising to visitors to find in the Great Valley region, and in other parts of the world, too, that the caves are under hills, and that before one can enter a cave he must climb up the side of the hill first. This is shown in the larger cross section.

There are many undeveloped caves in the Great Valley, and of course many sinkholes as well. One large sink, 1,000 feet in diameter, is immediately in front of the entrance to the Luray Caverns.

EXAMPLE 57   *The Problem*

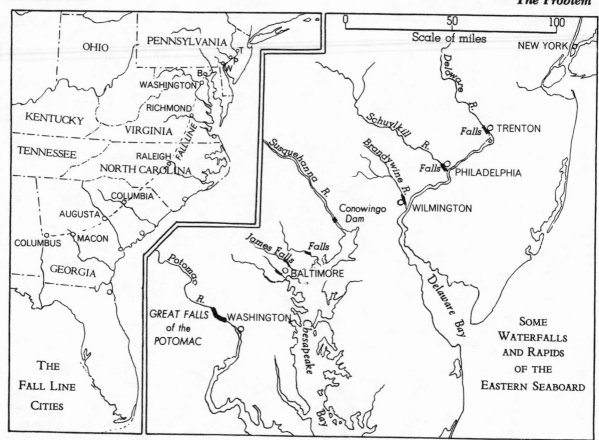

RIVERS. *Waterfalls. The Great Falls of the Potomac at Washington; the Schuylkill Falls at Philadelphia; the Fall Line.*

Just outside the city of Washington, a few miles to the west, is a picturesque region where the Potomac River comes foaming and dashing over a series of rapids and cascades. The so-called "Falls" here are of inconsiderable height, as is to be expected in a locality where the landscape is neither mountainous nor rugged. Rock Creek in Rock Creek Park likewise cuts its way through a narrow defile in another picturesque setting.

In Baltimore several of the streams passing through the city are confined to narrow gorges with rapids extending a mile or more along their courses. In these short stretches of rougher country, city and state parks have been established. The very names of the streams, such as Gwynns Falls, James Falls, and Stony Run which goes past Johns Hopkins University, suggest their character.

In Philadelphia, too, in the very heart of the city, is the well known Schuylkill Falls, now modified by a large dam. Passengers traveling on the Pennsylvania Railroad alongside the river can get a good look at them.

Also in Trenton, New Jersey, passengers glancing out of the window to the north as the train passes over the Delaware River can see the rapids in that stream which mark the upper limit of navigation for vessels coming up Delaware Bay from the ocean.

Richmond, too, has its waterfall on the James River which likewise is readily seen from the highways leading out of the city to the south. Almost all of the features mentioned above are to be observed on one or more of the familiar automobile road maps.

Still farther south, as shown on the above map, are other cities which stand on the so-called "Fall Line," in each one of which the streams have stretches of cascades and rapids, cities like Raleigh, Columbia, Augusta, Macon, and Columbus, Georgia. The problem is to explain the falls and rapids which occur along this "Fall Line" zone and almost nowhere else in the states south of New York.

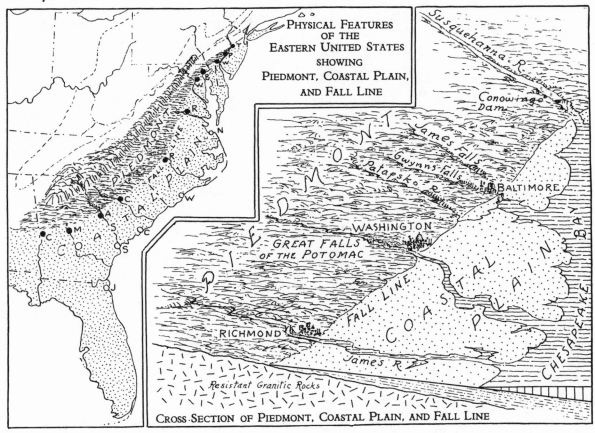

PHYSICAL FEATURES
OF THE
EASTERN UNITED STATES
SHOWING
PIEDMONT, COASTAL PLAIN,
AND FALL LINE

CROSS-SECTION OF PIEDMONT, COASTAL PLAIN, AND FALL LINE

The "Fall Line" cities of the eastern United States all lie along the boundary between the Piedmont area in the west and the Coastal Plain in the east. The Piedmont is a region of old crystalline granitic type of rock, whereas the Coastal Plain is made up of loose, unconsolidated layers of sand and clay. The Piedmont rocks are resistant to stream erosion, with the result that streams form narrow gorges with waterfalls and rapids where they are cutting into them. The Coastal Plain layers are weak. Here the streams widen out their valleys with ease, and their channels are not obstructed by obtruding ledges.

The "Fall Line" is not a simple sharp line but is rather a zone of considerable width. This means that the streams do not naturally have a single drop, although real falls result where the streams have been artificially dammed in these favorable places. This has been done with the Schuylkill at Philadelphia, the Susquehanna River at Conowingo Dam, and also on the James River at Richmond, Virginia.

Cities have come to be located along the fall line because the rapids in the streams have determined the upper limit of navigation from the sea. Even before the settlement of the United States by Europeans, Indian villages were established at these sites. These were the places where the Indians had to portage their canoes, and where naturally trading posts came into being.

The cities from Richmond north are not only "Fall Line" cities; they are also ocean ports because such streams as the James, the Potomac, the Susquehanna, and the Delaware have become "drowned" or submerged embayments. The "Fall Line" cities to the south, however, do not have this advantage. As a consequence they are much smaller cities. They have had to give way to such seacoast ports as Wilmington, North Carolina; Charleston, South Carolina; and Savannah, Georgia.

EXAMPLE 58   *The Problem*

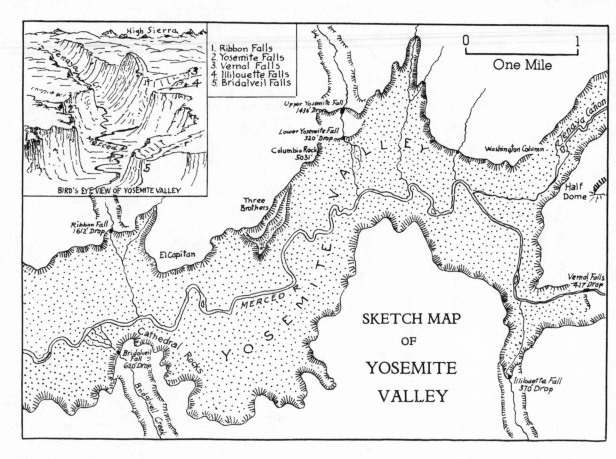

1. Ribbon Falls
2. Yosemite Falls
3. Vernal Falls
4. Illilouette Falls
5. Bridalveil Falls

BIRD'S EYE VIEW OF YOSEMITE VALLEY

Upper Yosemite Fall 1430' Drop

Lower Yosemite Fall 320' Drop

Columbia Rock 5031'

Three Brothers

Ribbon Fall 1612' Drop

El Capitan

Washington Column

Half Dome

Vernal Falls 717' Drop

SKETCH MAP OF YOSEMITE VALLEY

Cathedral Rocks

Bridalveil Fall 620' Drop

Bridalveil Creek

Illilouette Fall 370' Drop

MERCED RIVER

YOSEMITE VALLEY

Tenaya Canon

High Sierra

0   1
One Mile

RIVERS. *Waterfalls. Yosemite Falls, and Other Falls of Yosemite Valley.*

The above map is a simplified version of the small sketch map in the descriptive folder provided by the National Park Service for visitors to Yosemite National Park. The roads, trails, hotels, camps, and government buildings have been omitted. The names of the features have been reduced so as to emphasize a few of the significant details, especially the waterfalls. The waterfalls of the Yosemite are famous, not only because of their great height, but also for their beauty and their setting in one of the finest and most cherished gems of American scenery.

Waterfalls, wherever we find them, are unusual features of the landscape. They are indeed abnormalities. However, there are many kinds of waterfalls, just as there are many kinds of lakes. That is to say, there are many ways by which they have come about. The waterfalls of the Yosemite all belong in the same category inasmuch as they have all been formed in a similar manner. They are members of one family. To this same family group belong the waterfalls of the Alps, such as the Lauterbrunnen Falls, and those of the fiords of Norway and of Alaska, as well as the smaller falls of which the poets spoke in the "Lake Country" of England.

An examination of the above map shows that the falls of the Yosemite are hundreds of feet high, and most of them have a sheer vertical drop. Whereas Niagara Falls are only 140 feet high, Ribbon Falls of Yosemite and the Upper Yosemite Falls of Yosemite Creek are each more than ten times as much, higher indeed than the Empire State Building in New York City.

Yosemite Valley itself, drained by the Merced River, is a wide flat-floored affair rimmed by vertical walls which, because of their joints and cracks, have been carved into imposing architectural-like forms such as El Capitan, the Three Brothers, Washington Column, and the Cathedral Spires.

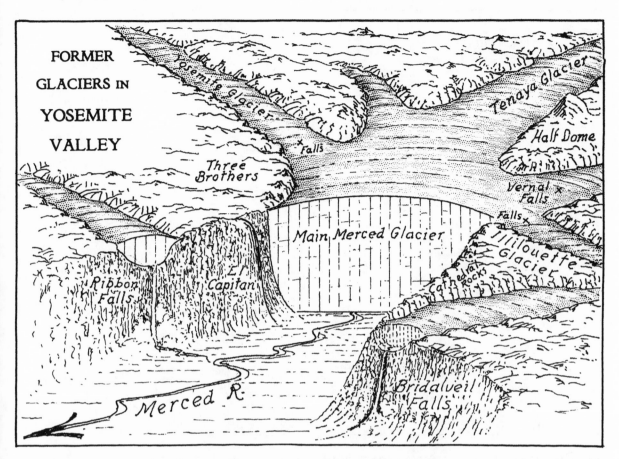

FORMER
GLACIERS IN
YOSEMITE
VALLEY

The large valleys and canyons which dissect the back slope of the Sierra Nevada, including Yosemite Valley, once supported an extensive system of glaciers. Like mighty rivers of ice, these glaciers flowed down from the snow fields of the High Sierra. Along this high crest small remnants of these original glaciers still exist. Pouring down toward the west, the glaciers came to an end in the lower country of the Great Valley of central California.

Having a thickness of several thousands of feet, the glaciers carved for themselves deep steep-walled U-shaped valleys or troughs, of which Yosemite Valley is an outstanding example. Tributary glaciers occupied the valleys where Yosemite Creek, Bridalveil Creek, Illilouette Creek, and other similar streams now flow.

It will be observed from the above diagram, at Ribbon Falls and Bridalveil Falls, that these tributary glaciers, being much smaller than the main glacier, cut valleys or channels of much less depth than that of the main glacier. This is the explanation for the many present waterfalls of the region. When the glaciers melted away from these valleys the smaller tributary valleys were left "hanging" above the main valley, the difference in elevation representing approximately the difference in thickness between the main and the tributary glaciers. As a matter of fact the glacier which once occupied Yosemite Valley was much thicker than the depth of the present valley suggests. The bottom of Yosemite Valley where the Merced River now flows has been filled with alluvium to a depth of many hundreds of feet. The result is that Yosemite Valley has a broad flat floor because the alluvium pretty much buries and conceals the rounded U-shaped form of the original glacial trough.

North of Yosemite Valley, but also within Yosemite National Park, is Hetch Hetchy Valley, somewhat smaller but just as beautiful as the famous Yosemite. It was, however, deprived of most of its charm when it was flooded by the Hetch Hetchy Reservoir to supply water for San Francisco.

EXAMPLE 59   *The Problem*

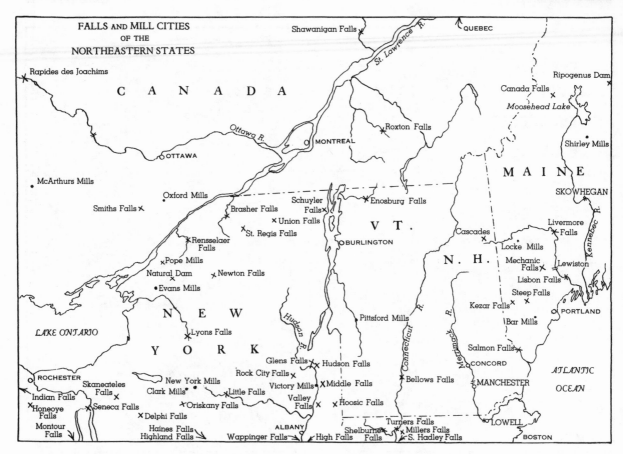

FALLS AND MILL CITIES
OF THE
NORTHEASTERN STATES

RIVERS. *Waterfalls. Some Falls and Mill Cities of the Northeastern United States.*

The above map, based upon the "Great Lakes Region" map of the National Geographic Society, shows a portion of New England, New York State, and Canada. On this map a *cross* indicates the location of towns having the name of "Falls," such as Glens Falls, New York. Also indicated is the location of towns having the name of "Mills," such as Pittsford Mills, Vermont. These are designated by a *dot*. A few other obvious rapids and dam sites are also indicated. A rough count of such places on this map reveals over three dozen towns bearing the name of "Falls" and ten more places called "Mills." Mills obviously were located in the earlier days where there were falls and consequently water power. This means about four dozen waterfalls of one sort or another in the area represented above. Doubtless many others have been omitted from the map.

Next, suppose we examine some of the more southern states, such as Virginia, West Virginia, and Maryland. For this purpose we will find one of the automobile road maps convenient, such as the Esso map that covers Virginia, West Virginia, Delaware, and Maryland. This map contains along its margin an Index of all the towns on the map. This makes it easy to find out how many towns bear the name of "Falls" or "Mills." In Maryland and Delaware there are no "Falls" whatever, and in Virginia and West Virginia only one each. The Virginia "Falls" is called simply "The Falls." It is situated on the Nottaway River and is clearly one of the "Fall Line" rapids, a type already described. In West Virginia the single town bearing the name of "Falls" is "Kanawha Falls" in the Kanawha River gorge, a feature doubtless caused by a resistant member of the Allegheny Plateau formations.

In the more northern New England states, many of the larger towns not bearing the name of "Falls" were originally located where there was abundant water power, such as Skowhegan, Auburn, and Lewiston, Maine; Manchester and Concord, New Hampshire; and Schaghticoke, New York. The number of falls and rapids in these states is clearly very great. The problem is to account for this.

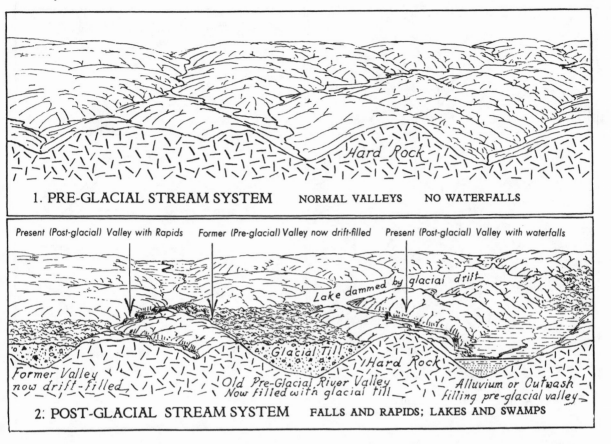

1. PRE-GLACIAL STREAM SYSTEM   NORMAL VALLEYS   NO WATERFALLS

Present (Post-glacial) Valley with Rapids   Former (Pre-glacial) Valley now drift-filled   Present (Post-glacial) Valley with waterfalls

Lake dammed by glacial drift

Glacial Till   Hard Rock

Former Valley now drift-filled   Old Pre-Glacial River Valley Now filled with glacial till   Alluvium or Outwash filling pre-glacial valley

2: POST-GLACIAL STREAM SYSTEM   FALLS AND RAPIDS; LAKES AND SWAMPS

The many waterfalls, rapids, and cascades, as well as the many lakes and marshes that occur throughout northern North America, owe their presence to the former continental glaciation of the region. Before the ice sheet spread over this area and temporarily obliterated all drainage lines, there existed what is called a "normal drainage pattern." Normal drainage is that which develops over a long period of time. Streams during this period widen out their valleys and erode their channels so as to do away with any irregularities caused by protruding rock ledges. For this reason waterfalls and rapids do not normally exist. This is true for most of the streams in the Southern states that were never covered by the ice sheet. The few exceptions to this can always be explained by some local aberration. The many lakes and swamps of the north occur in those portions of the stream valleys which were blocked by deposits of glacial "till" or what is also called glacial "drift."

Figure 1 above shows a hilly region with a normal stream system, prior to glaciation. The streams flow in well defined valleys without any interruption by falls and rapids. Nor are there any lakes or marshes. Many parts of the world are like this—regions where stream erosion has gone on uninterruptedly for long ages of time.

Figure 2 above shows the same region as Figure 1 as it appears after having been occupied by the ice sheet. The rivers have attempted to resume their former channels but now find that their valleys are blocked in many places by accumulations of glacial drift, moraines, and other kinds of deposits. Thus the streams are forced to find other courses. This means that in seeking a path through the hills, the streams find themselves flowing over rocky ledges. They consequently develop narrow gorges with frequent rapids and falls. These features, with the arrival of man, become the sites of mills and towns, and eventually of large cities, far too large in fact to be supplied by the meager water power which was adequate during the early days of the settlement.

EXAMPLE 60

The Problem

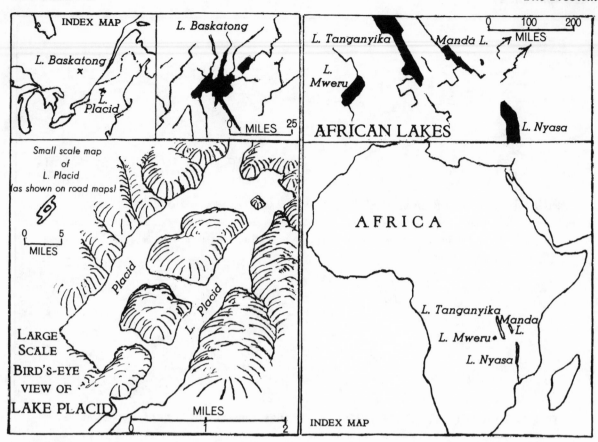

LAKES. *Rectangular Lakes. Lake Placid, New York; Baskatong Lake, Canada; Manda Lake, Tanganyika, Africa.*

Angular features in landscapes are unusual because nature rarely resorts to straight lines. And this is especially true of features depicted upon maps. Exceptions to this rule are therefore certain to attract attention. The examples shown above may readily be located upon easily available maps. Lake Placid is the best known of the three examples cited as it is visited by thousands of people each year, both in winter and in summer.

Lake Placid is situated in the Adirondack Mountains. It is shown clearly upon all automobile road maps, where it appears as in the small sketch map above. Close to Lake Placid is Mount Whiteface, whose summit is 3,000 feet above the surface of the lake. Motorists who drive to the top of Mount Whiteface can look directly down upon Lake Placid and observe its unusual shape. This kind of bird's-eye view of Lake Placid is pictured above. Most remarkable are the three rectangular islands in the lake, as well as the rectangular outline of the lake itself.

Another lake with rectangular characteristics occurs in the Province of Quebec, Canada. Baskatong Lake, however, is only one of the lakes in this vast lake-sprinkled region which exhibit straight-sided shores and right-angled branches. Maps on a scale of 1:2,000,000 or so are sufficiently detailed to reveal many of them.

Other continents, too, provide good examples, two or three in Africa being illustrated above. Not only the lakes but the rivers also have angular characteristics. Lake Urmia in northern Iran is still another instance. Like Lake Placid, it also contains an island which is an almost perfect square in shape.

While most straight-sided and rectangular lakes owe their form to some fundamental cause, nevertheless it is possible also that this shape in the case of some of them is merely fortuitous. In that case other features of the landscape may give a clue to the true explanation.

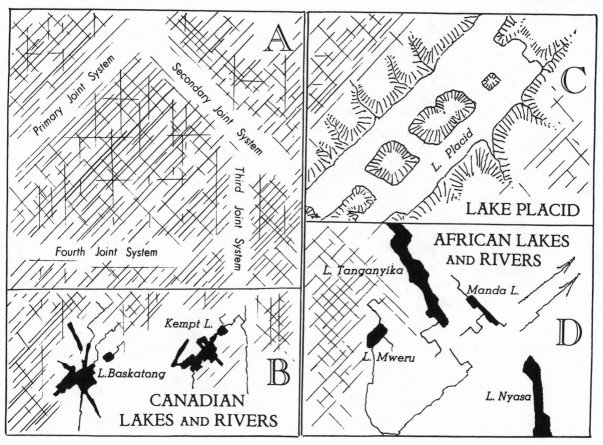

In many regions of the world, probably more frequently than is suspected, the rock formations are cracked or jointed. Inasmuch as the cracking is due to tremendous forces that are always acting upon the earth's crust, forces of gravity, mountain-making movements, tidal effects caused by the sun and the moon, as well as to the drying out and resulting contraction of formations, this cracking is apt to be widespread. Hundreds of miles of country are thus affected.

Moreover, this cracking has a regular pattern. The most common pattern is that produced by two joint systems which intersect more or less at right angles. There may be a third, a fourth, or even more sets of joints intersecting the first sets diagonally, as shown in Figure A above. The fact that this is brought about by pressure is in no sense theoretical inasmuch as it can be demonstrated in the laboratory where brittle material is subjected to constant strains from given directions. Similar rectangular joint patterns result.

The next point to be realized is the fact that joint lines are lines of weakness. It is along these lines and directions that drainage features develop. The rivers and lakes, therefore, assume rectangular forms. The Canadian lakes and rivers shown above correspond with the joint systems in a remarkable way.

Lake Placid is probably one of the most oddly shaped lakes we are apt to encounter anywhere. Its several rectangular islands are most unusual. It is easy to see, however, that all of these forms are related to a rectangular joint pattern, as shown in "C" above. Consult also the "Explanation" page for Example 44 where Lake Placid is again illustrated.

The African lakes and rivers have much greater dimensions than those of the two other areas mentioned. But here again the rectangular joint pattern is wonderfully reflected in the drainage lines, both large and small.

EXAMPLE 61   *The Problem*

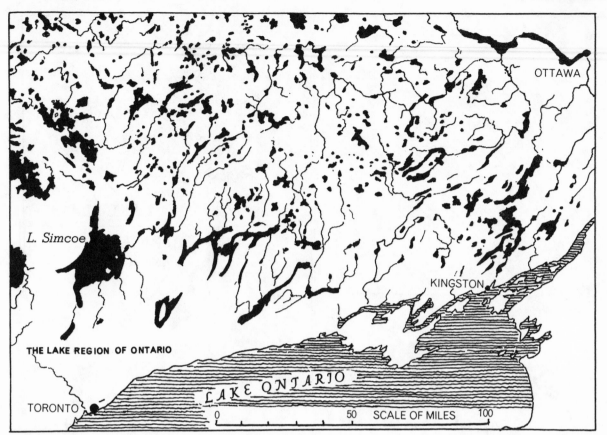

THE LAKE REGION OF ONTARIO

LAKES. *Lake Regions of the World. Northern North America and Europe, Represented by Ontario, Canada.*

Lakes of various kinds are scattered far and wide over the continents. But there are only two extensive regions where lakes occur in profusion and in a multitudinous variety of sizes and shapes. These two regions are, first, Canada, together with parts of the northern United States, and second, northern Europe, mainly Scandinavia and Finland. So familiar are most of us with lakes that we assume that lakes are common everywhere. But many parts of the world, and indeed of the United States, are almost devoid of them. West Virginia, for example, has no lakes. Nevertheless, Theodore Roosevelt when describing this part of the country in his historical work called *The Winning of the West*, remarks upon the forest-clad mountains "looking down perchance upon some almost-hidden lake." Even he, who had always lived in lake-dotted regions, unconsciously assumed that all regions must have lakes.

In some states with flat topography, like North Dakota, lakes are common, but other equally flat country like Texas is devoid of them. Some mountainous country like the Adirondacks has many lakes, but other quite similar mountainous country, like the Great Smokies, has none.

Spain has no lakes and France has virtually none. Though Canada has many lakes, Alaska has very few, nor does the vast area of Siberia have many. The great continents of Asia, Africa, and South America have no lake regions that are remotely comparable with those of North America and Europe. Nor does Australia have any, either.

The lake region of Ontario illustrated in the map above, like northern Minnesota, the "land of 10,000 lakes," resembles the lake country of Finland in many respects. Large and small lakes, long and short, very irregular at times, connected with each other by winding and twisting rivers with no apparent system—these are the characteristics of the two extensive lake regions of the world, those of North America and of Europe.

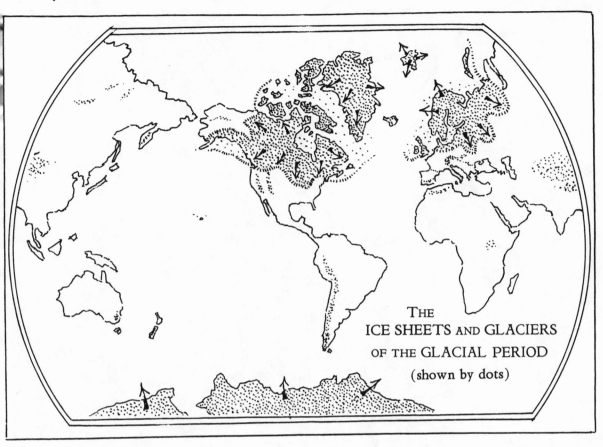

THE
ICE SHEETS AND GLACIERS
OF THE GLACIAL PERIOD
(shown by dots)

Most of the lakes in the northern latitudes of North America and Europe lie in those portions of these two continents that were covered at one time by the continental ice sheet. In the same way that the valleys of the Adirondacks were blocked by morainal debris to form such lakes as Lake Placid and Lake George, so also were the innumerable stream valleys of the north country blocked by debris from the melting ice to form the profusion of lakes that interrupt almost every water course. And equally abundant are the marsh lands that were brought about by the dislocation of the original drainage lines. Marshes, lakes, and irregular drainage lines, broken by many rapids and waterfalls—these are the characteristics of regions that have experienced glaciation.

The above map of the world shows the approximate limit of the last great ice sheets in North America and in Europe. In North America the ice spread out from several centers of accumulation in Canada, some of it even flowing northward. Note that in the Northwest Territory of Canada, in Yukon Territory, and in Alaska, the land was never inundated by the ice. And the same is true of Siberia. The coldest parts of the world, which were indeed amongst the driest ones, never received enough precipitation in the form of snow to form an ice sheet.

Except for the great ice sheet of Antarctica there was in the Southern Hemisphere no extensive glaciation. This is because there are no large land areas which reach sufficiently far to the south into the colder regions. In addition to the great continental ice sheets of North America and of Europe, there were also during glacial time many glaciers in the high mountain ranges of the world, such as the Rockies, the Andes, the Alps, the Pyrenees, the Caucasus, and the Himalayas. The Sierra Nevada of southern Spain and the Atlas Range of North Africa supported glacial systems too. Even in central Africa, right on the equator, on Mount Ruwenzori and Mount Kilimanjaro there was extensive glaciation. In all of these ranges and mountains small glaciers and snow fields still exist, mere remnants of the mighty rivers of ice that formerly flourished there.

EXAMPLE 62

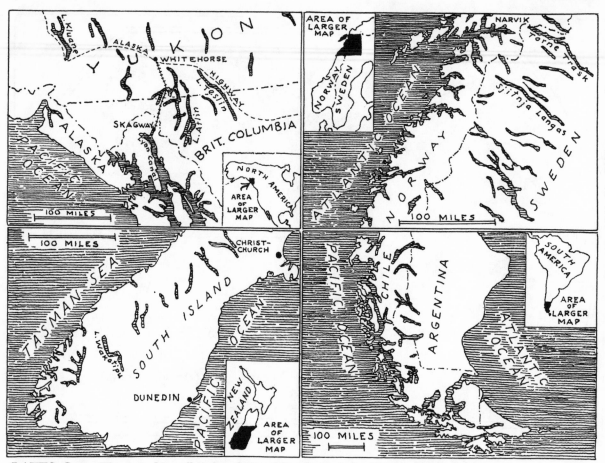

LAKES. *Some Finger-Lake and Related Fiord Regions of the World. Northern Canada, Sweden, New Zealand, and Chile.*

The four regions shown above constitute the four most important finger-lake regions of the world. There are of course other finger-lake regions, such as those of Switzerland and northern Italy, as well as that in Glacier Park in the United States. But nowhere do finger lakes occur in such size and in such numbers as they do in the four localities that we are considering here.

There is another reason, however, for presenting these four regions. It is to call attention to the fact that these are also the four most important fiord regions of the world. We have already seen in Example No. 22 that these fiord regions lie in the two belts of prevailing westerly winds, one in each hemisphere. Is there any connection also between these wind belts and the finger-lake regions? Or, in other words, why should finger lakes and fiords be so numerous in the same general localities?

The great fiord region of the Alaska and British Columbia coast has as its equivalent in the interior the finger-lake region of northern Canada. Still farther south, in British Columbia, is a continuation of this finger-lake region, just north of the United States border.

The fiords of Norway are balanced in the same way in the interior of Scandinavia by the remarkable finger lakes of Sweden. These lakes are actually longer and almost more impressive than the fiords themselves.

In New Zealand likewise the fiords of the coast have their counterpart in the finger lakes of the interior. These wonderful lakes are not so well known by the people of the Northern Hemisphere, although they constitute one of the main attractions of New Zealand.

Then there are the renowned finger lakes of Chile and Argentina, a region now becoming famous as a resort area, not unlike Switzerland in several respects. These lakes, like the other three regions discussed here, also have their counterpart in a coast line replete with fiords. One of these lakes lies far to the south near the tip end of the continent in the land of Tierra del Fuego.

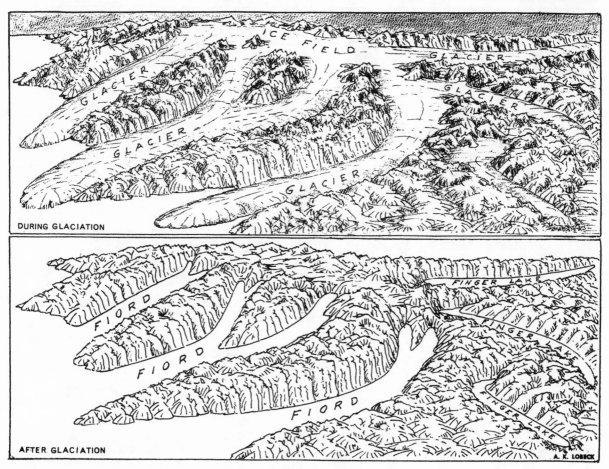

The similar origin of finger lakes and fiords is depicted in the above illustration. Both finger lakes and fiords, as we have already seen, are the result of glacial action. The only difference between the two is that on one side of the mountain crest, that is the western side in all of the cases mentioned, the glaciers reached the sea. On the other side of the mountain crest, however, the glaciers died out as they reached lower levels. In each of the four regions mentioned, small glaciers, or as a matter of fact some quite large ones, still reach tidewater. In Alaska, for example, at the head of Glacier Bay, several large glaciers, such as Muir Glacier, come right to the coast. In New Zealand the well known Franz Joseph Glacier comes very close to the coast where it comes down to the lower valleys of the heavily forested mountain range.

Of the important fiord regions of the world Greenland alone does not have any corresponding finger-lake region. This is merely because of the fact that the whole interior of Greenland is still covered by a great ice sheet. The fiorded coast of Greenland has many glaciers which reach sea level and there break off to form great icebergs which float southward even as far as Newfoundland, where they become a menace to trans-Atlantic shipping.

In the four areas discussed on the preceding page, the mountain crest was covered during Glacial Time by a vast snow field. This snow field or névé behaved in some respects very much like a continental ice sheet. The snow and ice poured outward from the mountain crest to the east and to the west, as shown on the top illustration above. Similar ice fields covered much of the Rocky Mountain region during Glacial Time, constituting what is called the Cordilleran Glacier. Most high mountain regions, during Glacial Time, supported ice fields of this type. The tongue-like glaciers emerging from these fields scoured out troughs that later became finger lakes. Such are the finger lakes of the Alps and the Rocky Mountains. Nevertheless, the greatest finger lakes of the world correspond with the mountain areas that had the greatest ice fields. These were the four regions that received the heaviest precipitation, the regions lying close to the coast on the western side of the several continents.

EXAMPLE 63

The Problem

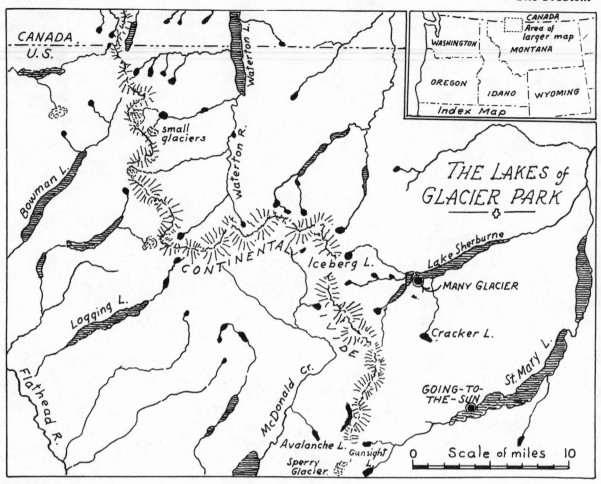

LAKES. *Lake Swarms. Some High Mountain Lakes. Glacier Park.*

The above map was sketched from one of the popular folders of Glacier Park issued by the National Park Service. The roads and trails and many names have been omitted in order to bring out more clearly the lakes of this region. There are two types of lakes shown here. Most conspicuous are the long finger lakes, such as Lake Sherburne and St. Mary Lake. But most numerous are the very small circular lakes. These occur in the high mountain area at the very heads of the streams. This type of lake is known as a "tarn." Along the Continental Divide are the highest peaks of this mountain area.

The finger lakes, which are shaded on the map, occupy former glacial troughs or valleys. The glaciers in Glacier Park at one time were vastly larger and longer than they are at present. They extended both east and west from the crest of the range which is now the Continental Divide separating the Atlantic from the Pacific drainage systems. At the present time only small remnants of these former large glaciers now exist. Sperry Glacier, near the bottom edge of the map, is the one most frequently visited by tourists. These small glaciers occur now only in the highest part of the range.

The location of the small lakes or tarns is noteworthy. Shown in solid black on the map, they occur at the very heads of most of the rivers. In fact, they occur in the same location with relation to the streams as the small glaciers do.

The tarns are remarkably small. Not one of those shown on the map is as much as a mile across. Most of them are round. A few are slightly elongated. The problem here is to explain the presence of the numerous small round lakes in contrast with the long finger lakes.

Although lakes of this type are common in the high mountain ranges of the world, none of them appear on any of the usual atlas maps because of their small size.

130

A MOUNTAIN PEAK with several Glaciers forming glacial cirques at their heads

MOUNTAIN PEAK after melting of glaciers. Tarns occupy the cirques.

The above sketch shows two mountain peaks along a high divide similar to that in Glacier Park. The peak on the left is being glaciated. Several glaciers, like rivers, are moving outward from its summit. At their heads the glaciers, unlike streams, gouge out cup-like hollows in the mountain mass. These hollows are known as glacial amphitheaters or glacial "cirques." In their early stages glacial cirques are actually quite circular. This is illustrated on the rounded mountain mass in the far distance. As the cirques become larger they encroach upon the whole mountain summit so that eventually only the core of the mountain remains. This core resembles a pyramid. Many are sharp-pointed like those illustrated above. Such peaks are known in geology as "Matterhorn Peaks," named after Mount Matterhorn, which is a typical example. Most glaciated mountain ranges have peaks of this kind. This gives the range a serrate or saw-tooth form. From this comes the Spanish term "Sierra" which is applied to many high ranges. The term Sierra Nevada, for example, means the "Snowy Saw-Toothed Range."

At the right, in the illustration above, is a second peak, similar to the other one. The glaciers, however, have melted away, leaving their rounded glacial troughs or valleys. The cirques at the head of the troughs now contain small lakes or tarns. These lie in the rock basins that have been scoured out by the ice. There may be several lakes in a chain, as is shown in the middle of the three troughs.

Tarns are always *small* lakes. Rarely are they as much as a mile in diameter. They are among the most picturesque lakes in the world. Towering crags rise above them, sometimes several thousand feet, to the mountain summit. Small glaciers may still be found at the heads of the tarns. These occasionally break off into the water to form icebergs. This accounts for the name "Iceberg Lake" which has been applied to one of the lakes in Glacier Park, shown on the map opposite. Gunsight Lake, Cracker Lake, and Avalanche Lake are other examples, all well known to Glacier Park visitors because of their accessibility by trail.

EXAMPLE 64   *The Problem*

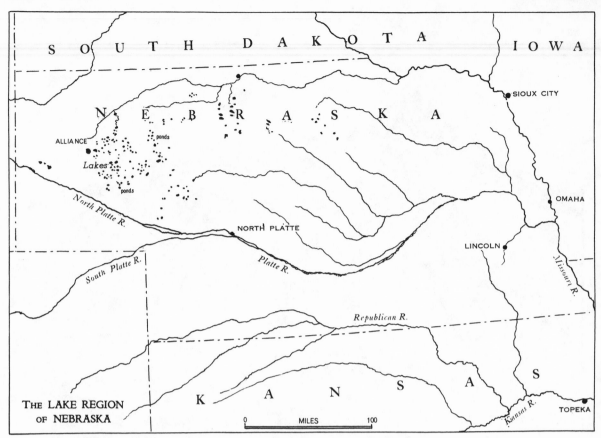

THE LAKE REGION
OF NEBRASKA

LAKES. *Lake Swarms. Some Dry-Region Lakes. Lakes of Western Nebraska.*

In western Nebraska on the High Plains is a swarm of small lakes and ponds. There are many hundreds of them, perhaps even thousands. On the average map most of them are mere dots. The automobile map of Nebraska shows several score of the largest ones. Most of these larger ones bear names. Several of them are water-fowl sanctuaries. They all lie off the beaten path and are visited by few people outside Nebraska. The transcontinental motorist crossing Nebraska will almost surely use U.S. Route 30, which follows the valley of the Platte River. If in western Nebraska he turns north to the small city of Alliance, and then again turns east on State Road No. 2, he will find himself in the heart of the lake area.

The remarkable thing about this swarm of lakes is its uniqueness. That is to say, it is the only swarm of lakes within the vast limits of the High Plains. The above map indicates clearly how the lakes occur within a limited area. Eastern and southern Nebraska is quite devoid of them. Kansas has none, South Dakota has none, nor does Colorado.

We have already noted that many lake areas lie within glaciated country. For several reasons we may be sure that the lakes and ponds of Nebraska were not caused by continental glaciation. In the first place, they lie south of the area covered by the continental ice sheet. In the second place, the lakes are not shaped like glacial lakes. They are mostly round and quite uniform in shape, without the numerous irregularities that glacial lakes have. Not only that, but they are not connected with any river systems. The lakes in fact do not drain out. About all we can infer from the map is that each pond occupies a little depression or hollow of some kind. The rainfall of the region is apparently not enough to fill up these hollows and cause them to overflow. Why this lake swarm should be restricted to so limited an area is hard to understand. In other dry areas of the world lakes of this type may also be found, serving as sites of oases in desert regions.

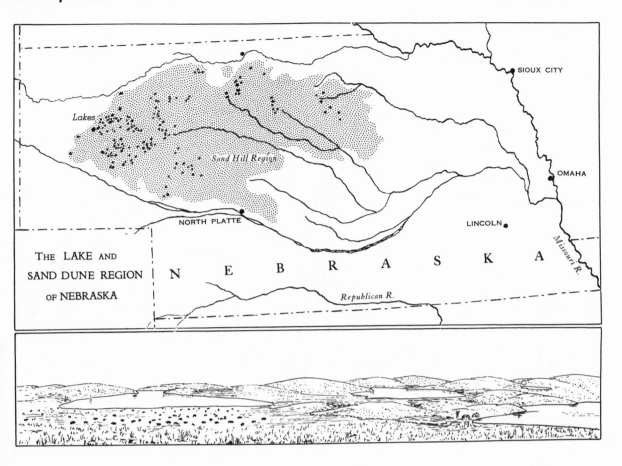

THE LAKE AND SAND DUNE REGION OF NEBRASKA

The lakes and ponds of western Nebraska are indeed unique. They all occur within the limits of the area known as the "Sand Hill Region" of Nebraska. The area covered by these sand hills is delineated on the above map. It embraces some 24,000 square miles of country, which is one-half the size of the entire state of New York. The hills which cover this region are all dunes, that is, sand dunes blown up by the wind. The lakes and ponds occupy the hollows between the dunes. Over several thousand square miles the dunes are still drifting; elsewhere they are more or less fixed. The area of drifting sand varies with the rainfall from season to season. In rainy summers most of the area is well grassed and supports a considerable cattle industry. In many of the hollows ponds no longer exist. Ground water, however, is not far beneath the surface. It is secured by means of windmills, which are a common adjunct to almost every ranch in the Great Plains.

The visitor first viewing the Sand Hill Region would believe that it is totally uninhabited. But soon he discovers that almost every hollow contains a ranch house of some type. And at intervals he would see large herds of cattle grazing in the meadows close to one of the ponds. Each pond may average only ten acres or so in area. The small sketch above is an attempt to present a typical picture of the region.

Many of the lakes contain brine and potash. In fact, in some years the output of potash from brine lakes of western Nebraska is equal to one-third of the total output of the United States.

The reason for the formation of dunes within this limited portion of the High Plains is explained largely by the character of the soil. The soil here has a high percentage of sand, which is readily drifted by the wind and piled up as dunes. Farther east in Nebraska the soil consists of "loess," that is, a fine wind-blown dust. This wind-blown material, however, rarely assumes a dune-like form.

EXAMPLE 65

The Problem

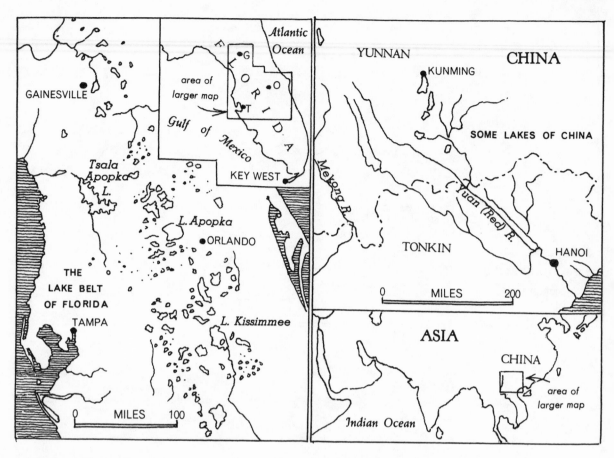

LAKES. *Lake Swarms. Lakes of Florida; Lakes of China.*

In northern Florida, in the region lying between Lakeland, Orlando, and Gainesville, there is a profusion of lakes. Here also are some of the springs for which Florida is famous, such as Silver Springs and Juniper Springs. This region of lakes and springs forms a belt which extends from close to the northern boundary of the state southward for over two hundred miles to the northern edge of the Everglades, not far from Lake Okeechobee. Here the lakes end abruptly. Lake Okeechobee, however, is not a member of the Florida Lake Belt.

Most of the Florida lakes have fairly simple outlines. Some, however, like Lake Tsala Apopka, have the most irregular-looking shore lines imaginable. On atlas maps only a few of the larger lakes are shown, but on large-scale maps like the automobile road maps the lakes appear in untold numbers. Many of these are mere ponds. Some of the lakes have outlets, but customarily they lack drainage, either coming in or going out.

Southern Florida is entirely devoid of lakes except for Lake Okeechobee, which is almost unique in its class. In the far western end of Florida is another small colony of lakes which is similiar in origin to those which are here under consideration. These two lake areas of Florida, although they lie far beyond the southernmost limit of continental glaciation, nevertheless on most maps resemble closely the lake swarms of northern regions.

Another lake region, similar in origin to those of Florida, occurs in the Yunnan Plateau of southern China. Here are many small basins, many of which contain lakes and ponds. Most maps of China that are readily available to us, are on a scale too small to show these lakes in their full profusion. Most maps, however, will show the largest of these lakes, Lake Tien Chih. This lake stands on the floor of the Kunming Basin, or Kunming Plain, upon whose floor many other smaller but similar lakes also occur.

None of the maps, from which the above sketch maps were made, suggests in any way how these lakes have come to be there.

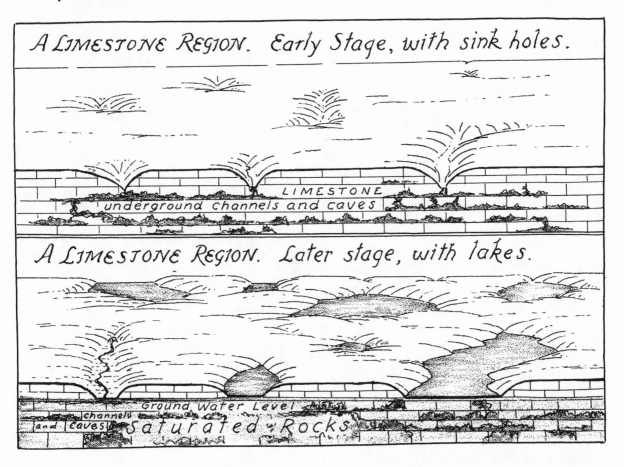

The lakes of Florida lie in a region of very soluble limestones. In a limestone region the rain which falls upon the ground seeps into the rocks, through cracks and joints, and creates beneath the surface extensive passageways and underground channels. The larger of these become caves. In the uppermost figure above, there is shown an early stage in this development. The surface of the ground becomes marked by numerous funnel-shaped depressions, or "sinkholes." Into these depressions the streams of the region "sink" or disappear. Gradually the sinkholes become larger and larger. Some of them fuse with each other until most of the country consists of nothing but depressions, both large and small. If the land stands at a relatively high level above the sea, the water which goes underground works its way along until it eventually reaches a river. But if the ground is relatively low-lying and also fairly level, as it is in Florida, and the climate is rainy to boot, much of the water fails to flow off. The accumulation of water underground, the so-called "ground water," gradually rises to such a height that it fills the bottoms of the sinkholes. This is shown in the lower one of the two figures above. Many lakes are thus produced. In places where the underground water gushes out near the side or bottom of a sink, springs result. Some springs are veritable rivers, where the underground streams emerge into the open.

Unlike Florida, the Yunnan Plateau of China is a high region. It is a limestone area pitted with many depressions or sinks which are occasionally known as "dolines." It is a true karstland like the famous Karst region of Yugoslavia, along the eastern Adriatic. Under such conditions many of the depressions are dry open basins.

In this connection see also Example 50 which describes the formation of sinks and other Karstland phenomena.

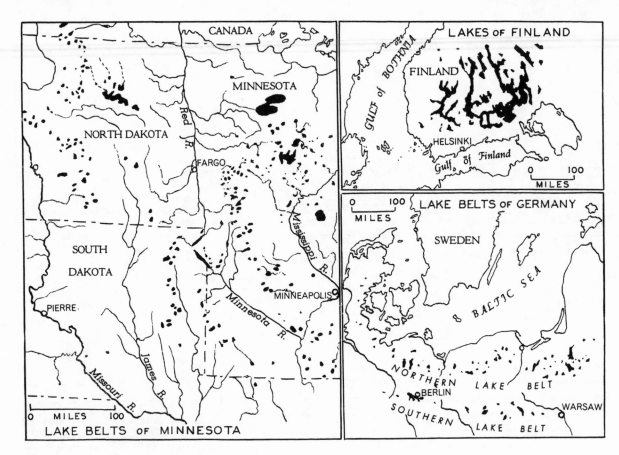

LAKES. *Lake Belts. Minnesota; Finland; Germany.*

In this example attention is directed to two or three regions selected from the United States and from Europe. They are not unique, for similar examples can be found elsewhere in the world. In their larger aspects these regions are relatively flat. They are in fact known as "plains." That these regions are endowed with numerous lakes is well known. Minnesota, for example, advertises itself as the "Land of 10,000 Lakes." This is probably no exaggeration if all the thousands of small ponds are included. Finland, too, is known as the land of lakes. The lakes of Finland are so interconnected that one can go by water, with very little if any portaging, between all parts of the country. In the lake belt of Germany, too, the region of former East Prussia is still known as the Mazurian Lake District, and the area lying between Berlin and the Baltic Sea is called the "Pomeranian Lake Region." The lakes in all of these regions are obviously the most notable features of the landscape.

One aspect of these lake areas, however, is not so fully appreciated; namely, the fact that the lakes are arranged more or less in belts, and that these belts are roughly parallel. Perhaps we might ask, "How else could they be arranged?" They could of course be scattered hit and miss, without any system, or they could be more or less uniformly spread over the area like the many small ponds of western Nebraska. Or they could be like the lakes of Maine, some large, some small, some very irregular, some elongated, and all hidden among the many hills and mountains to be found uniformly everywhere throughout the state.

The lake belts of Minnesota, it may be observed, resemble fairly closely the lake belt of central Florida. On large-scale topographic maps the lakes of the two areas seem similar. The resemblance, however, is only superficial, and careful study reveals a number of important and characteristic differences. Other features of the landscape, too, bear this out.

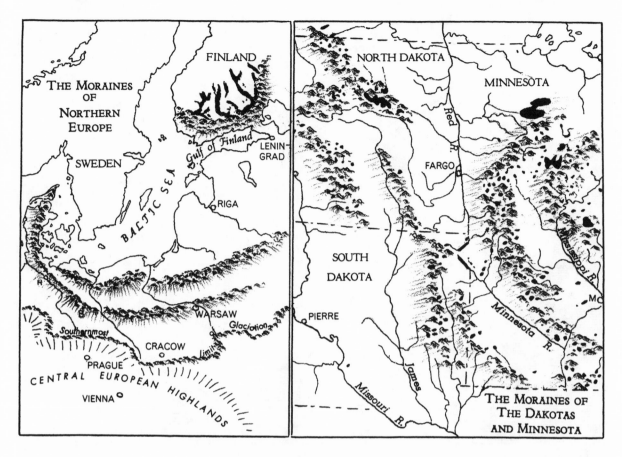

Lake belts result from several causes. In Florida the lakes, as we have seen in a previous example, occur in belts of limestone country. Lakes occur frequently along river courses, being parts of the old river bed. They occur as lagoons along many coasts. The lakes of Minnesota and northern Europe, however, occur in the morainal belts, that is the terminal moraines left by the continental ice sheets. The terminal moraines are belts of small hills and intervening hollows or depressions. Many of these depressions are dry because the loose porous soil of which the moraine is formed will not hold water. Others, especially the deeper ones which go below the water table, contain lakes. Many of the lakes are intermittent, disappearing during dry seasons. Others gradually become clogged with vegetation and change to swamps or peat bogs. Some represent merely blocked-up portions of river valleys that have been obstructed by the moraine.

Look at the regions under consideration, taking Finland first. The lakes of Finland are mostly drowned river courses that have been dammed by the moraine. The moraine, as shown on the above map, forms an arc around the southern end of Finland. The arc shape conforms with the lobate form of the ice sheet. This moraine was probably left during a period of halting as the ice sheet was melting back after it had reached south to the region of the North German plains.

The lakes of North Germany occur in two fairly well defined belts conforming with two locations of the terminal moraine. The curved lobate form of these moraines is clearly shown on the above map.

On the map of Minnesota shown above several moraines are depicted. The elaborate moraine system of Minnesota and the Dakotas suggests the many oscillations of the ice front during the period of ice retreat. Similar belts of moraines and lakes occur in Wisconsin and in northern Illinois, Indiana, and Ohio.

EXAMPLE 67   *The Problem*

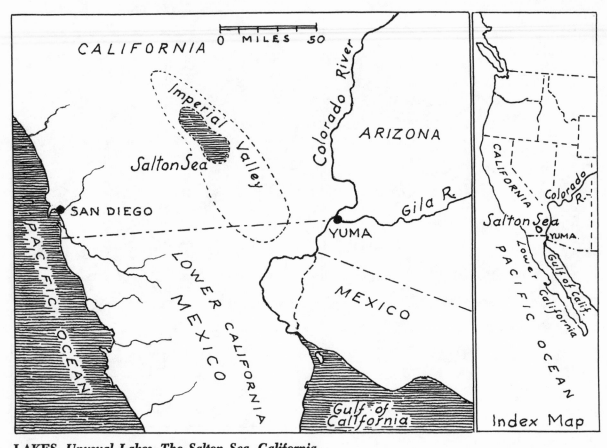

LAKES. *Unusual Lakes. The Salton Sea, California.*

In southern California, close to the Mexican border, is the famous Imperial Valley. In the center of this valley is Salton Sea, a salt lake whose surface is 240 feet below sea level. The lake, therefore, has no outlet to the ocean. Were this a more humid region the lake basin would fill up and overflow into the Gulf of California. On some occasions in the past the Colorado River nearby has broken out of its main channel and has discharged vast volumes of water into the Imperial Valley. This has greatly increased the volume and the area of Salton Sea. The Southern Pacific Railroad at one time ran close to the shore of this lake, but had to be relocated when the lake spread out and covered a greater area. Gradually the lake subsided again because of the excessive evaporation in this dry climate.

Like most lakes occupying closed basins, that is, basins which they do not overflow, Salton Sea is a salt lake. In that respect it resembles Great Salt Lake and other lakes of the Great Basin in Utah and Nevada. The reason for this is that in this dry and desert-like climate the relatively small amount of water which flows into the lake is insufficient to fill up the lake basin. The water draining into Salton Sea carries a minute trace of salt in solution, as does all water which drains from the land. Because this salt never has a chance to get out of the lake, it accumulates from year to year and becomes more and more concentrated. In Great Salt Lake, for example, there is more than seven times as much salt, for a given volume, as there is in the ocean. Some salt lakes have even more than that.

It seems strange indeed that this valley should have become one of the most important agricultural regions of California, or for that matter of the United States. This has been accomplished through irrigation. Here the conditions for irrigation are ideal. Water can be drawn from the Colorado River through canals and led into the basin, which is everywhere below the level of the Colorado. Thus, with abundant water, constant sunshine during almost every day of the year, and unusually fertile soil—as is the case with most desert soils—phenomenal crops can be achieved.

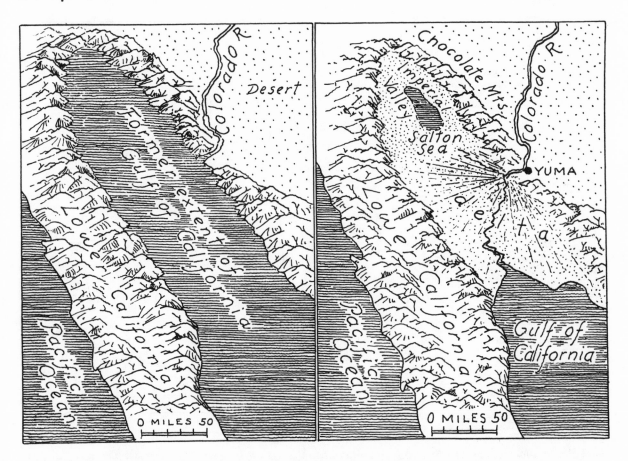

The two illustrations above show how the Imperial Valley was cut off from the Gulf of California by the building of a delta by the Colorado River. In the illustration at the left is portrayed the former extent of the Gulf, which reached much farther north than it does at present. It occupied a long trough-like depression between the mountain ranges. Then gradually the Colorado River, bringing great quantities of silt down from the Rocky Mountains and from the Colorado Plateau through which it was cutting its great canyon, proceeded to build a delta into the quiet waters of the Gulf. There were no currents which might sweep this material away into deeper water. We have seen in Example No. 2 that many streams have built deltas into the relatively quiet waters of the Gulf of Mexico, where alongshore currents are less powerful than in the open ocean. So also here, as in most protected bays and gulfs, the conditions for delta building were favorable.

There are in the world many other examples of this sort, where deltas have cut off part of the sea to form lakes. The case of Salton Sea differs from most of the other cases, however, in the fact that the head of the gulf which was cut off evaporated and left a valley instead of a large sea-level lake. This valley—the Imperial Valley—is merely the floor of the old gulf, and Salton Lake is the last remnant of the upper gulf.

The illustration above is somewhat simplified, for as a matter of fact there are several other salt lakes in this region. One of them lies to the south within the limits of Mexico.

Among the best known examples of so-called "delta lakes," formed by the cutting off of part of the sea by a delta, is Lake Ponchartrain near New Orleans, produced by the building of the Mississippi Delta. Zuyder Zee in Holland was formed by the building of the Rhine Delta.

In a more or less similar manner the Nile delta separates the Red Sea from the Mediterranean Sea.

EXAMPLE 68

The Problem

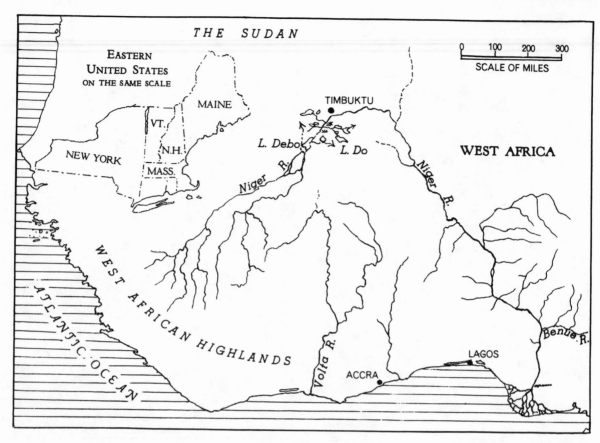

THE SUDAN

EASTERN
UNITED STATES
ON THE SAME SCALE

MAINE

VT.

N.H.

NEW YORK

MASS.

L. Debo

TIMBUKTU

L. Do

Niger R.

Niger R.

WEST AFRICA

WEST AFRICAN HIGHLANDS

ATLANTIC OCEAN

Volta R.

ACCRA

LAGOS

Benue R.

0   100   200   300
SCALE OF MILES

LAKES. *Shallow Lakes and Swamps. Four Great Swamplands of Africa. The Timbuktu Region; Lake Chad Region; the Sudd of the Nile; and the Okavango Swamp.*

Ordinarily we do not associate swamps with deserts, or even with semi-arid country. But in Africa there occur, at the very edge of the greatest deserts of the world, four extensive swamplands. In each of the swamps there are many shallow lakes as well. These several regions are each some 200 miles across.

These regions, too, are all contiguous to the largest rivers of Africa, the Niger, the Nile, and the Zambezi. Three of these regions lie north of the equator, along the southern margin of the Sahara Desert. This is the belt known as the Sudan. We may term these northern regions the "Timbuktu Region," the "Chad Region," and the "Sudd." The Timbuktu region lies along the course of the Niger. The Chad region is on the Logone, which occasionally runs into the Niger; and the Sudd lies along the Nile. South of the equator, the Okavango Swamp lies on the Okavango and other tributaries of the Zambezi. See the map of Africa on the opposite page.

Into the Timbuktu region, as shown on the map above, flow the Niger River and other large streams from the mountainous country to the southwest. There the Niger seems to break up into a number of smaller channels which diverge in several directions and end in a multitude of small lakes and swamps. Finally the Niger manages to emerge from this network of lakes and streams and pursues its way southeastwardly to the sea.

Lake Chad receives drainage from numerous streams rising in the higher country to the south. It is surrounded by a vast swamp which is flooded at times by the lake itself, and on those occasions this enlarged lake overflows through the Bahr el Ghazal to the even larger Kiri Depression to the northeast.

The Sudd is a vast swampy region into which flow many streams from the southern uplands, including the Nile itself, or rather that portion known as the White Nile, which comes out from Lake Albert.

Into the Okavango Swamp, which occupies the northern part of the Kalahari Desert, flow several large rivers from equatorial Africa. There, in the desert, these streams break up into many distributaries. At times of high water the Okavango emerges from this maze and continues eastward to join the Zambezi near Victoria Falls.

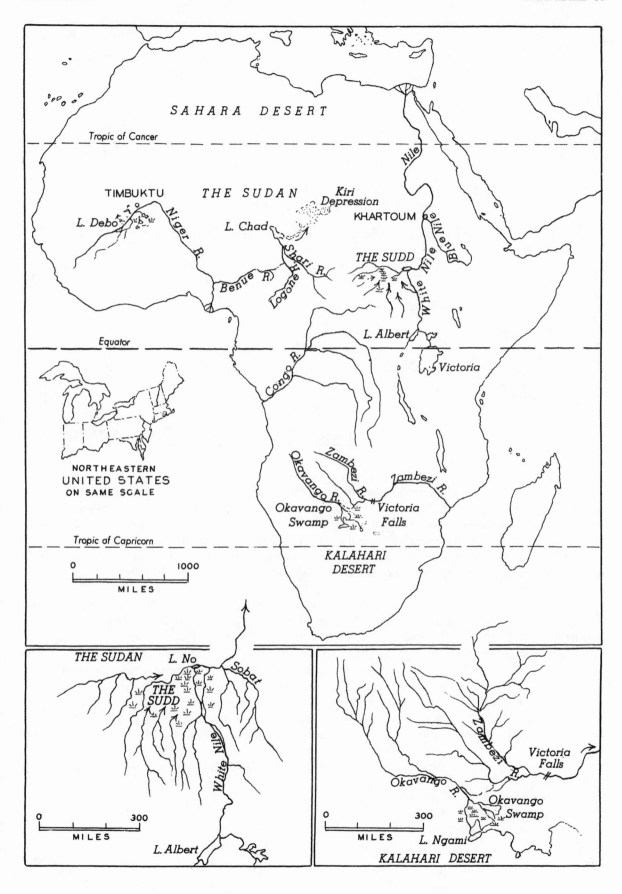

EXAMPLE 68   *The Explanation*

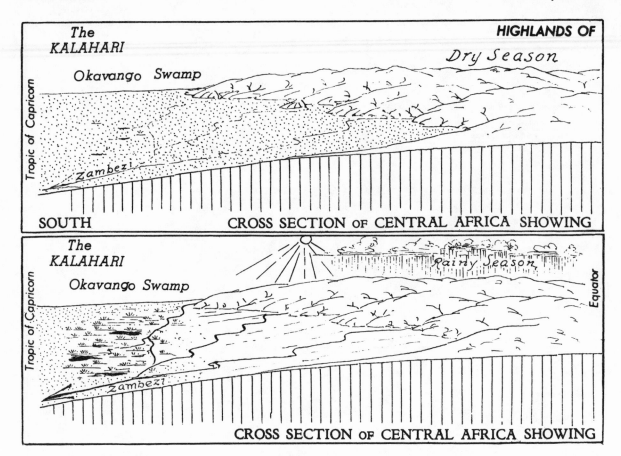

The four great lake and marshland regions of Africa change mightily in area from season to season. During the wet season the lakes expand tremendously because of their shallow depth. Then the marshes with their interlacing watercourses become a network of reed-grown channels, an intricate maze that baffles all but those who are familiar with their every detail. During the dry season the lakes shrink to a mere fraction of their former size. The marshes then dry up and become a prey to vast grass-fires set by lightning or more frequently by the natives. The animal life, hippos and crocodiles, concentrate in the few waterholes that remain, waiting for the wet season to return.

Let us examine this whole situation a little more closely by means of the above diagrams. These diagrams, it will be noticed, extend across the two pages. They represent a fanciful cross section of central Africa extending lengthwise from the Tropic of Capricorn on the south to the Tropic of Cancer on the north, north being at the right. The upper part of the diagram shows the situation as it exists during the northern summer, between June and September. At this season the sun is north of the equator. In this whole equatorial region the rainy season comes at the time that the sun is most nearly overhead, which is the summer. This meteorological fact comes about because the great heat of the sun causes upward-rising convectional currents in the atmosphere. These rising currents, upon reaching the higher levels of the atmosphere, expand. Cooling results from this expansion and this brings on rapid condensation, causing great cumulus clouds and much rain, often of the thunderstorm type.

In the northern summer the rains fall north of the equator. In the southern summer the rains fall south of the equator.

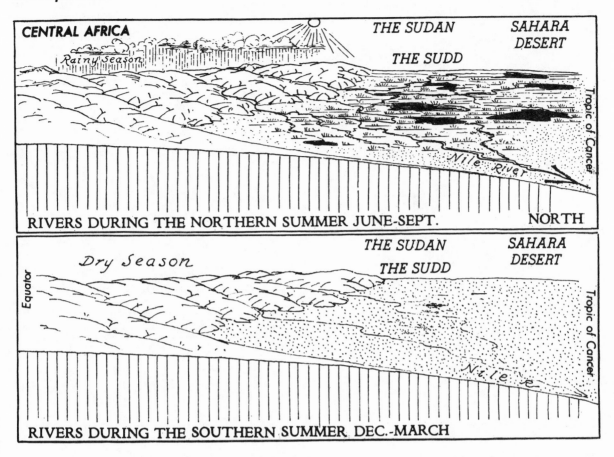

The headwater portions of the rivers, during the rainy season, receive vast volumes of water, more indeed than streams like the Nile or the Niger can accommodate. These many tributaries of the larger rivers, and many other streams too, bifurcate and flow off in all directions into the surrounding desert. Extensive alluvial flats are thus formed which, while the rainy season lasts, become almost like big inland seas.

As the sun returns to the southern sky, the northern rivers tend to dry up, the marshes and lakes decrease in area, and the major rivers, like the Nile, dwindle in volume. The southern rivers then become flooded. The Okavango and other streams pour great volumes of water into the Okavango Swamp, and some of their many channels find their way to the Zambezi. At this season, too, between December and March, the Victoria Falls on the Zambezi have a stupendous volume and are truly an awe-inspiring sight. This is the time to see them, near the end of the southern rainy season.

From these statements it is obvious that these several marshlands alternate between periods of abundant and periods of deficient water. Irrigation, therefore, is carried on. Much of the land in the Timbuktu region has in that manner been brought into cultivation by French engineers. It is easy to see, therefore, why this fertile and rich tract of country became in early times the terminus of one of the most important caravan routes across the Sahara.

In South Africa studies are now being carried on in the Okavango region to determine the practicability of bringing much of this area into cultivation by irrigation methods. Professor John K. Wellington, of the University of the Witwatersrand, recently in that connection visited the United States to study our irrigation practices, notably in the Sacramento delta area of California.

EXAMPLE 69

The Problem

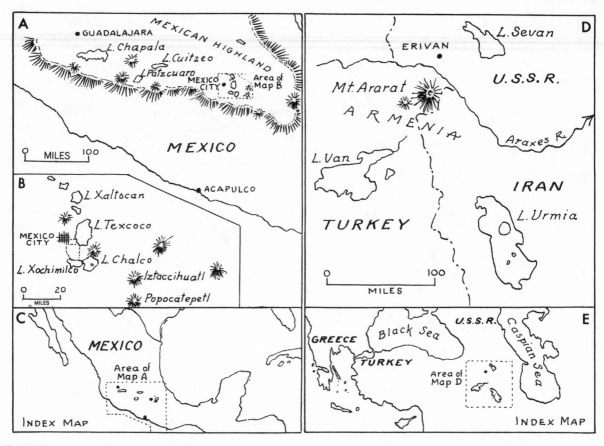

LAKES. *Some Volcanic-Region Lakes. The Lakes of Mexico; the Lakes of Armenia.*

On the high plateau of Mexico, at elevations of 5,000 to 6,000 feet above sea level, there are several unusual lakes, including those near Mexico City. These are shown, on different scales, on three of the above maps. Maps A and B are redrawn in a much simplified manner from a large German atlas called the *Stieler Hand-Atlas.* Map C is an Index Map to show the general location of the lakes. These maps are used in preference to more modern maps because they show the several lakes near Mexico City, most of which have since become dry lake beds.

In the days of the Aztecs, before the destruction of Montezuma's capital city of Tenochtitlán, now Mexico City, by Cortez, in 1520, the intermontane basin or Valley of Mexico, called also the Vale of Anáhuac, was covered by several bodies of water. The largest of these was Lake Texcoco, and only slightly smaller was Lake Chalco, both shown on Map B. At that period a network of canals, Venetian in their multiplicity, covered the neighborhood and connected the many villages along the lake shores with the central square of the capital city where there stood a huge pyramid.

Floating gardens, originally made of interlacing twigs like a thick mat and covered with earth, almost concealed the lakes. From these gardens the Indians in their primitive dugouts carried produce to various parts of this bizarre island empire. Of all the lakes now only Xochimilco remains, largely a maze of tranquil canals bordered by floating flower gardens, and a source of interest to all visitors.

Among the larger lakes of the Mexican plateau are Lake Chapala, Lake Pátzcuaro, and Lake Cuitzeo. All of these lie in intermontane basins hemmed in by volcanoes and, like the lakes of the Valley of Mexico, are shallow bodies of water, the banks grown up in reeds.

Another region having lakes similar to those of the Mexican Highland lies in what was formerly known as Armenia, now part of Turkey, Iran, and the U.S.S.R. (See Figures D and E above.) Here are three large shallow lakes, Lake Van, Lake Urmia, and Lake Sevan, all about equidistant from the beautiful cone of Mount Ararat.

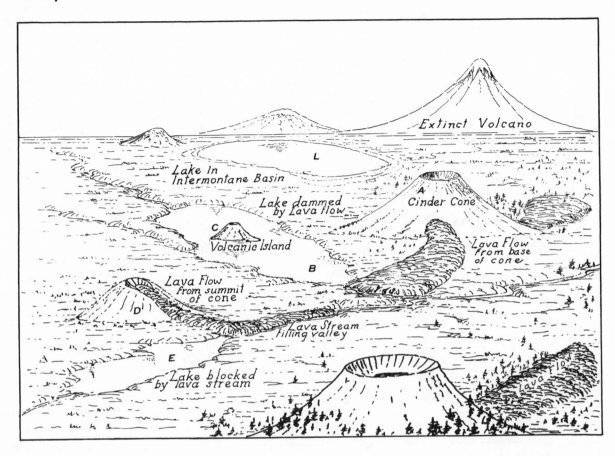

In volcanic regions like central Mexico, and like Armenia, the irregular distribution of volcanoes tends to form intermontane basins. These naturally become occupied by lakes which spread out between the volcanic cones. Such volcanic cones explain also the occasional occurence of circular islands in the lakes themselves.

Another important cause of lakes is the damming of river valleys by lava flows. An example well known to visitors to the National Parks is Snag Lake near Lassen Peak in Lassen National Park. Several other similar lakes lie in its vicinity. These all appear on the sketch map which accompanies the Park Service booklet provided for visitors.

A region similar to the Mexican Highlands is depicted on the above sketch. "L" is a lake occupying a basin area surrounded by volcanoes. "A" is a cinder cone or small volcano perhaps a thousand feet or less in height. It is built of loose fragments of material ejected through its crater. Near its base, on this side, is a recent lava flow which has blocked up a valley to form Lake "B." This lake completely surrounds the small cinder cone "C" which forms a circular island. From another small volcano, "D," there is also a lava flow which has run down a valley, almost like a stream of water. This lava flow has also caused the formation of a lake, Lake "E."

On some lava plateaus, like the Columbia Plateau of Washington and Oregon, extensive lakes have been formed as the result of lava flows. These lakes, now extinct, have left far-flung expanses of silt or lake deposits. Interbedded among the lava flows which make up the plateau are still earlier lake beds. These lake deposits are now important sources of underground water. Where they outcrop along the walls of canyons, springs and waterfalls result.

The lakes of Armenia, such as Lake Van and Lake Urmia, do not have any outlets, and are consequently salt lakes. Being shallow also, their area changes from season to season. Lake Chapala in Mexico is also a very shallow lake. The name Chapala, or Chapalal, is a Nahuatl Indian word in onomatopoeic imitation of the sound of waves lapping on the beach.

EXAMPLE 70   *The Problem*

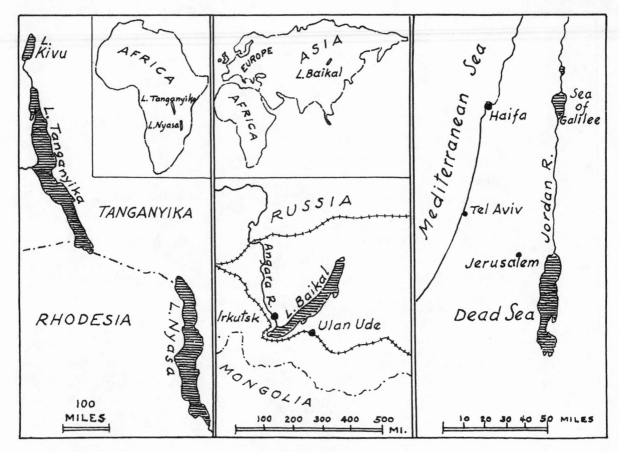

LAKES. *Some Very Deep Lakes. Tanganyika, Nyasa, Baikal, the Dead Sea.*

The group of four lakes presented here includes examples of some of the deepest lakes in the world. Lake Baikal in central Asia is in fact the deepest known lake. It is much more than a mile in depth at one point, and its bottom is 4,226 feet below the level of the sea. Its total length is almost 400 miles, its width about 50, giving it therefore an area of 20,000 square miles. This large area, combined with its great depth, causes this lake to contain a greater volume of water than any other fresh-water lake in the world. Lake Tanganyika in Africa, with a depth of 4,700 feet, is the second deepest fresh-water lake in the world, its bottom being also far below sea level. Lake Nyasa, like both Lake Baikal and Lake Tanganyika, is a long and relatively narrow lake, and likewise a very deep one.

Another very unusual lake that belongs also in this category is the Dead Sea. It happens to be a salt lake, but this is an accident of climate, because this is one of the driest spots in the world. The Dead Sea is a remarkable lake in that its level stands over a thousand feet below sea level, and its bottom is 1,300 feet below that, a remarkable depth indeed for a lake of that relatively small size, for it is less than 50 miles in length and only 10 miles wide.

Even more remarkable than the Dead Sea is Lake Tahoe in California, between Nevada and California. This small lake is only 21 miles long and 10 miles wide, but it has a depth of over 1,600 feet, which is considerably deeper than the Dead Sea. The origin of these very deep lakes is one of the interesting stories of geology.

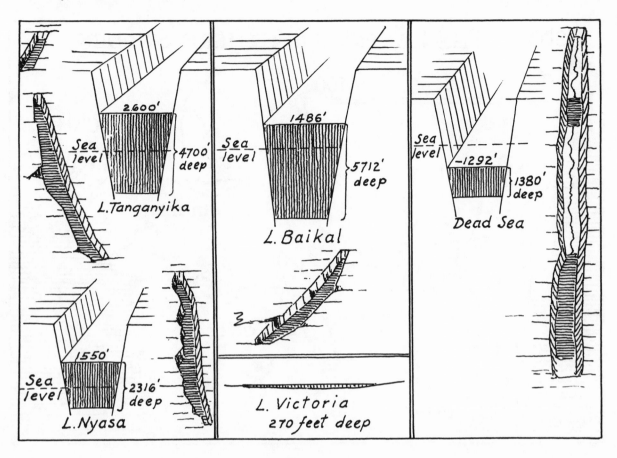

Tanganyika, Nyasa, and several other African lakes, as well as Baikal, the Dead Sea, and Lake Tahoe, all occupy parts of long trench-like depressions, known as "rift valleys." Rift valleys are known also as "grabens," a German word meaning a "grave" or ditch. These grabens or rifts are formed by a fracturing of the earth's crust along two parallel planes. The long slab lying between the two fractures or "faults" was dropped down below the rest of the country to form a long depression or rift. The longest and deepest rifts in the world occur in East Africa. The westernmost of the two largest rifts contains Lakes Nyasa and Tanganyika, as well as Lakes Edward and Albert farther north. The eastern rift contains a number of salt lakes such as Natron, and Naivasha near Nairobi. Naivasha indeed for many years served as the landing area for the flying boats of the B.O.A.C. coming from England. The northern end of the eastern rift runs all the way up to the northern end of the Red Sea, where it forms the Gulf of Aqaba and the long deep valley that holds the Dead Sea and the Jordan River.

The cross sections above, together with the diagrammatic sketches, show the trench-like form of the several lake valleys just mentioned. The cross section of Lake Victoria is presented also to show how shallow it is, as it is only 270 feet in depth. It is, however, a very large lake, being second only to Lake Superior in area.

Not all rift valleys contain lakes. One of the best known grabens is Death Valley in California, whose floor is 280 feet below sea level. Lying in the Great Basin, just east of the Sierra Nevada, it has a very dry climate. Evaporation here exceeds the rainfall.

EXAMPLE 71   *The Problem*

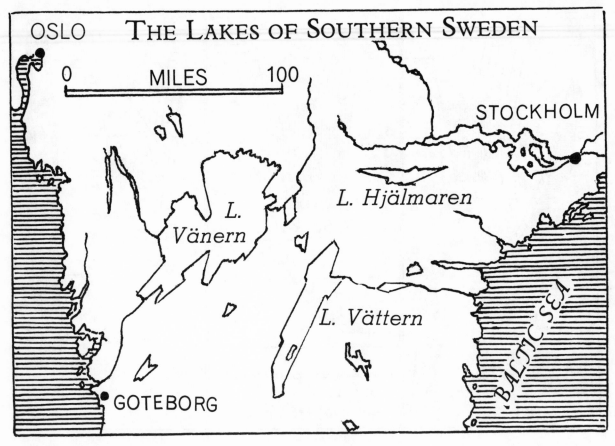

LAKES. *Some Straight-Sided Lakes. Lakes of Southern Sweden.*

In southern Sweden are two large lakes, Vänern and Vättern, and many smaller ones. Their angular outlines, even on an ordinary map, are fairly noticeable. Although these lakes lie in a glaciated region, their shape is the result of a totally different cause. Although these lakes have large areas, each one being close to 100 miles long, they are relatively shallow bodies of water, being only 300 to 400 feet deep at most. Lake Vänern is 90 miles long and 321 feet deep; Vättern is 80 miles long and 390 feet deep. Not all of the lakes of southern Sweden are straight-sided and angular like Vänern and Vättern. Some are long and snake-like. These long lakes are merely parts of river valleys which have been blocked by glacial deposits. The larger lakes obviously occupy broad lowlands of some kind. The problem here is to explain the presence of these lowlands and to account also for their angular shape. Like the other lakes of Sweden, they have been blocked by glacial deposits, but the lake basins themselves have not been formed by ice action.

Connecting several of the lakes is a remarkable canal system which makes it possible to cross Sweden by boat from Goteborg to the Baltic Sea.

The sketch map shown above is enlarged to about twice the size of the average atlas map of Sweden in order to show better the details of the region.

The lakes of southern Sweden resemble in several respects those of the Adirondack Mountains. Angular patterns are characteristic of each group. But as we shall see, the Swedish lakes involve a somewhat different kind of history. We might note also that the lakes of southern Sweden are very different from those of Finland described in Example 66.

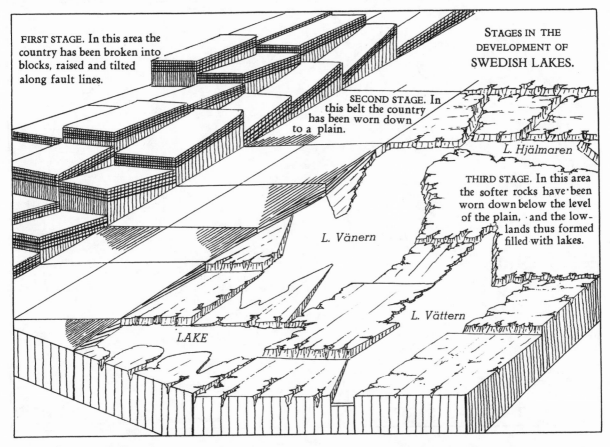

FIRST STAGE. In this area the country has been broken into blocks, raised and tilted along fault lines.

STAGES IN THE DEVELOPMENT OF SWEDISH LAKES.

SECOND STAGE. In this belt the country has been worn down to a plain.

L. Hjälmaren

THIRD STAGE. In this area the softer rocks have been worn down below the level of the plain, and the lowlands thus formed filled with lakes.

L. Vänern

L. Vättern

LAKE

A study of the lakes of southern Sweden reveals not only their angular outlines, but field investigation shows also that the lakes occupy depressions or lowlands from which weak sedimentary beds have been removed by erosion. The surrounding steep-sided plateau-like areas are made up of more resistant rock. It remained for a distinguished American geologist, Professor W. M. Davis, to demonstrate how this condition was brought about. The above diagram, modeled after one of Professor Davis's drawings, illustrates this in three stages.

The first stage is shown in the back part of the drawing. Here we see a region broken into blocks. These blocks have been elevated to various heights, and some have been tilted in various directions. It will be noticed that the top portion of each block consists of some sedimentary layers of rock. Before the country was broken up, these sedimentary beds formed a continuous layer across the country. Beneath these layers is a different kind of rock. This foundation rock is a resistant crystalline rock, such as granite, upon which the sedimentary beds have been deposited.

In the second stage, in a belt just this side of the block area, we see the country after the blocks have been eroded down to a plain, to what geologists term a "peneplain." This was accomplished over a long period of time during which the streams wore everything down virtually to sea level. This plain bevels across crystalline and sedimentary rocks alike, the sedimentary exposures being shown in the above illustration by the shaded patches.

In the third and final stage we see the same region after erosion has again attacked it. This renewed erosion was brought about by the bodily uplift of southern Sweden above the sea. The rejuvenated streams have naturally removed the weaker more easily eroded beds first, thus forming the lowlands. The more resistant masses remain as plateaus or as sloping uplands. As a final step the region was glaciated, and the valleys and lowlands were blocked with glacial deposits, so that lakes resulted.

EXAMPLE 72   *The Problem*

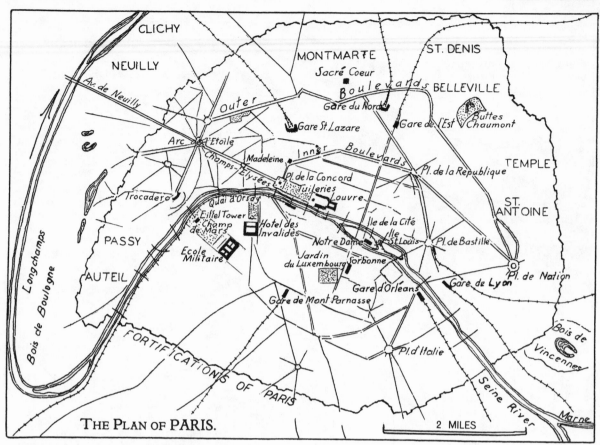

THE PLAN OF PARIS.   2 MILES

CITIES. *Circular Cities. The City Plan of Paris.*

The plan of any city, its streets, its railroads, its parks and open spaces, its distinctive buildings reflect two sets of controlling circumstances. First, there is the topography of the region. This includes both the immediate local topography and also the broader topography of the country as a whole. In other words, the plan of a city is influenced by its regional setting. In the second place, a city reflects the conditions of the historic times in which it exists and during which it has developed. Paris serves as an admirable example to illustrate both of these factors. But first of all, on this page we shall try to get some notion of what Paris looks like on a map.

Paris lies athwart the course of the Seine River. This stream cuts the city into two parts as it swings in great meanders from its junction with the Marne River just east of the city gates. In the heart of the city the Seine flows around two islands, the Ile St. Louis and the Ile de la Cité upon which stands Notre Dame.

Surrounding Paris is the city wall or "enceinte," a continuous series of now dismantled ramparts or fortifications, outside which is a moat or depression. Constructed in 1840, it withstood the siege of 1870–1871. It is pierced by numerous openings, gateways or "portes" for the highways and railroads that enter the city.

The "boulevards" of Paris form a series of concentric circles. These are fairly discernible on the map above. The so-called "Inner Boulevards" enclose the old medieval Paris with its irregular streets. The Inner Boulevards, sometimes called the "Grand Boulevards," run from the Church of La Madeleine around to the Place de la Bastille. Within this circle is the real heart of Paris, the Louvre, the Tuileries, the Opéra, the Place de la Concord, and the Cathedral of Notre Dame.

In the outer part of the city, beyond the Outer Boulevards but within the city wall, are the Faubourgs or suburbs, some of which are industrial and others strictly residential in character.

Many modern avenues and plazas now cut through the older part of Paris, avenues which intersect upon some "circle," "square," "place" or "étoile," exhibiting the same radial symmetry so beautifully exhibited in our own French-inspired city of Washington.

150

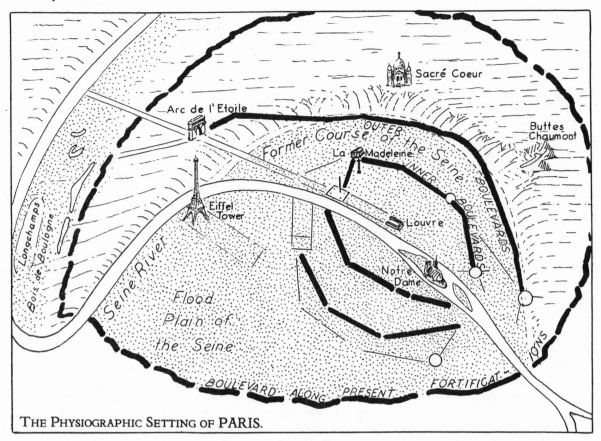

THE PHYSIOGRAPHIC SETTING OF PARIS.

Paris stands in the center of the Paris Basin. It is thus the focal point for all of northern France, one-third of the whole country. Fifteen railroad lines and an equal number of national roads converge upon it. In Example 47 we noted the centripetal arrangement of the rivers also. A large city was destined to grow up in just about this locality.

The more immediate cause for the location of Paris at this point was the presence of two islands in the Seine River. Settlements on these islands could be more readily defended than' elsewhere. Following the early European tribes, the Romans occupied this site and maintained it for several centuries. The defensive wall built around the margin of the Ile de la Cité was the first of the concentric walls surrounding the city of Paris. Later walls were built in 1020, 1180, 1370, and in 1625, each successive wall forming a larger circle surrounding the others. The 1625 wall, built by Louis XIII, determined the belt of the Inner Boulevards. The word "boulevard" comes from the German word *Bolwerk*, or bulwark, meaning a rampart, because the boulevards were built on top of or alongside the ramparts.

The next ring encircling the city, the sixth of the series, is now represented by the ring of the Outer Boulevards. This ring, it will be seen from the diagram above, lies largely along the foot of the circular belt of heights surrounding Paris on the north. It represents the location of the old meander scarp of the Seine River. This wall, being at the foot of the cliffs, was not for defensive purposes, but was designed to keep the farmers from entering the city to sell their produce without paying the proper tax or duty. The Buttes Chaumont forming part of the surrounding heights were formerly a quarry supplying building stone for Paris. Other heights are represented by the hill of Montmartre where the Church of Sacré Coeur stands. Likewise, the Arch of Triumph and the Trocadero occupy a further continuation of this high belt.

It will be seen from this that much of Paris occupies the flood plain of the Seine River. Old meander scars and tributary streams have been obliterated. However, in the Bois de Boulogne, several abandoned meanders of the Seine River form linear winding lakes in the park.

MORE UNANSWERED PROBLEMS

From our perusal of this book we must conclude that the features depicted upon maps must all have some explanation. Here only a few of these things have been touched upon. Many more similar problems come to mind. The continents, themselves, why are they there? And all the rivers of the world, how did they come to be where they are? And then there are the man-made details, the boundaries of states and countries, the location of cities, the roads and railroads, and even the names on the map—there is a reason back of all of these facts. Some are more noteworthy than others. Why does the boundary between Maine and New Hampshire depart a little from a true north-south line? Why is the boundary of northern Delaware a circle? Why is there a little piece of Minnesota 'way up across the Lake of the Woods, entirely detached from the United States, thus being the northernmost point in the country? And the boundaries of Europe. They have had a fantastic history, one which is still going on. From our maps as we studied them in school we would think these boundaries were fixed and permanent, but now we are beginning to know better. And the cities and towns of the world, many very ancient, many very new, why are they where we find them? And why have some of the ancient cities ceased to be?

And the names on the map, too, their origin is fascinating beyond words. Why so many cities bearing classical names in upstate New York—Rome, Utica, Troy, Syracuse, Ithaca, Seneca? Whence comes the name Coblenz in Germany? It derives from the Latin *confluens*, at the confluence of the Moselle and the Rhine. And the aboriginal names in Australia: Warrambool, Wangaratta, Katoomba, Toowoomba, Wagga Wagga— these, like the Indian words used in the United States that are equally strange-sounding, all mean something in the native tongue. All over the world people of the past have contributed the names we use at the present. For instance, the Spanish heritage of our Southwest is preserved in our Sierra Nevada, Arizona, Colorado, canyon, Mesa Verde, Rio Grande. The Aztec names in Mexico such as Popocatepetl, Iztaccihuatl, and Xochimilco, the French names in Louisiana, the Dutch names in New York and New Jersey, and of course the well known English names in New England, all of these record some facts of historical significance. The Chinese names, too, so strange to us, have a real meaning, a knowledge of which in many cases helps us to understand their locations.

Maps not only tell us a great deal; they raise even more questions than they answer. A map is simply loaded with problems to excite our curiosity. A map is a window not only to the present features of the earth's surface, but to the past as well. Constantly we ask the questions: How did these things come to be? What was their past origin?

There is still another category of things appearing on maps, features and details in nature which are strange or curious. We find a number of things ascribed to the Devil, such as the Devil's Tower, the Devil's Punch Bowl, the Devil's Post Pile, the Devil's Slide, Devil's Lake, and so on. There is also Hell's Canyon and Hell's Half-Acre. There are the Badlands and, even worse, Death Valley and the Dead Sea. But we find also the Garden of the Gods, Paradise Valley, and the Mountain of the Holy Cross. These names, fanciful though they are, are designed to be descriptive of the localities to which they are applied. Have we ever wondered what these places really are? The Devil's Post Pile: what are these so-called "posts" and how did they get there? In short, what is the real explanation for all of these curiosities of nature? I have sometimes visited caves and listened to the guide pointing out the interesting features to the tourists, the "kissing camels," the "beehive," the "bacon strips," the "seal," the "throne room," the "bottomless pit," and other similar attractions. This is interesting and diverting, but not nearly so stimulating as a discourse on how the various shapes came about. Cave formations, like the larger features of the landscape, have come about by natural processes, and once visitors understand how these odd shapes have originated they view them with an even greater interest and appreciation.

Let me mention some more places of geological interest which are rarely understood by the visitor, and let me append too a little statement concerning their real meaning. To be awed merely by the fantastic aspects of a scene, without any deeper understanding, is to my mind only a form of shallow satisfaction unworthy of any real lover of nature.

The "Hoodoos" of Yellowstone Park, near Mammoth Hot Springs (a mass of great jumbled house-sized blocks of rock resulting from a landslide, where in stagecoach days hold-ups were common).

The "Hanging Hills" of Meriden, Connecticut (trap ridges arranged en echelon because of a series of diagonal faults).

"Shiprock," New Mexico, resembling a great sailing ship at sea (actually the neck of a large eroded volcano).

"Liberty Cap" in Yosemite Park (one of the great granite "domes" of the Sierra upland, cut by a joint face).

"Rock City," near Olean, New York (like other similar "rock cities" in the Allegheny Plateau, a massive bed of conglomerate cut by joints that have widened to produce the so-called "streets").

"Valley of 10,000 Smokes" (fumaroles left by the destruction of Mount Katmai volcano).

The "Kaiserstuhl" (a volcanic neck situated along the fault line marking the eastern side of the Rhine Graben).

The "Giant's Causeway" (a cross section formed by wave erosion of an area of basaltic columns).

The "Côte d'Or," the "Côtes de Meuse," and so on (these so-called "coasts" are limestone scarps once believed to have been formed by the waves, like the white cliffs of Dover, but now known to have been formed by ordinary weathering and stream erosion).

The Needle's Eye, the Witches Cauldron, Monument Valley, the Great White Throne, Elephant Butte, Book Cliffs, Gunsight Pass, these and many other places seen by the Western traveler all embody some fact of geological interest. They should be considered not merely as curiosities but as striking examples of geological phenomena.

INDEX

Places or features shown only on the maps are not listed.
Boldface figures, e.g. **67**, indicate EXPLANATIONS.
Italics, e.g. *21*, indicate *maps or illustrations*, in addition to text references.

Montauk Point, Long Island, **54–55**
Monte Argentario, Italy, **35**
Montmartre, Paris, **151**
Moraines: Denmark, 7; Long Island, 55; Minnesota, Finland, and Germany, 137; New York region, 47
Mount Adams, 94–95; Mount Hood, 94–95; Mount Matterhorn, 131; Mount Mazama, 95; Mount Pelée, 67; Mount Rainier, 94–95; Mount Washington, 95
Mud springs, 114
Muir Glacier, Alaska, 129
Muskeget Island, Nantucket, **56–57**

Naivasha Lake, Africa, 147
Nantucket, **56–57**, 79
"Narrows," New York, **46–47, 55**
Natural levees, 89
Nauru Island, Pacific, 69
Nebraska, lakes, **132–133**; parallel rivers, **86–87**
Neufchâtel, France, 85
Nevada, interrupted rivers, **106–107**
New Guinea, 2, 64
New Hebrides, 64
New Jersey, barrier islands, **40–41**
New York Bay, **22–23, 46–47, 55**
New Zealand, 14–15, 64; finger lakes, **128–129**; geysers, 114
Newark Bay, **22–23**
Ngauruhoe volcano, New Zealand, 15
Niger River, 140
Nile River, 140
Nittany Valley, Pennsylvania, 91
No Mans Land, Martha's Vineyard, **56–57**
Norris Geyser Basin, Yellowstone Park, **115**
North America, lake regions, **126–127**
North Dakota, moraines, 137
North Downs, England, 29, 49
North Mountain, Nova Scotia, **23**
North Sea, 16–17
North Star, in latitude determination, 3
Norway, fiord coast, **44–45**
Notre Dame, Paris, 150–151
Nova Scotia, 23
Nyasa Lake, Africa, **146–147**

Oahu Island, Hawaii, **72–73**
Ocean Island, 69
Oceanic islands, **68–69**
Oder River, Germany, 7
Offshore bar, **38–39**
Ohio River, **80–81**
Okavango swamp, Africa, **140–143**
Öland, Island, Baltic, **52–53**
Oldland, 75
Onega Lake, Russia, 19
Ontario, Lake, 18–19
Ontario lake region, **126–127**
Orient Point, Long Island, **54–55**
Ortelius, world map of 1589, 2

Ouse River, England, **28–29**
Outer lowland, 78–79
Outwash, Great Plains, 87
Outwash plain: Cape Cod and Martha's Vineyard, 57; Denmark, 7; Long Island, 55
Oxford, location, 29

Pacific Ocean: atolls, **68–71**; islands, **64–65**; Ortelius Map of 1589, 2
Padre Island, Texas, **38–39**
Pago Island, Dalmatian coast, 62
Palisades of the Hudson, 23, 25, 26, 27
Parallel rivers, **84–85, 86–87**
Parasitic spits, 33
Paris, location, 98; city plan, 150–151
Paris Basin, 99
Pays de Bray, France, 85
Pearl Harbor, location, 73
Peddocks Island, Boston Bay, **58–59**
Pelée, Mount, 67
Peneplain, 149
Peninsulas: Banks, New Zealand, 14–15; Bayonne, New Jersey, **22–23**; Brittany, **16–17**; Cornwall, **16–17**; Cotentin, France, **16–17**; Croton Point, **24–25**; Denmark, **6–7**; Door, Wisconsin, **18–19**; Florida, **8–9**; Hel, Poland, **30–31**; Kent, England, **16–17**; Keweenaw Point, **20–21**; Korea, **10–11**; Saugeen, Ontario, **18–19**; Taranaki, New Zealand, **14–15**; Tunisia, **12–13**
Penn Valley, Pennsylvania, 91
Pennsylvania: caves, 116; dendritic rivers, **102–103**; trellis rivers, **90–91**
Philippine Islands, 64
Phoenix Islands, 64
Phosphate, in Florida, 9
Piedmont, eastern United States, 9, **119**
Pikes Peak, 94–95
Pindus Mountains, 13
Piracy, stream, 111
Pitch, of fold, 63
Place names, 152
Placid, Lake, **124–125**
Plain, outwash, 7, 55, 57
Playas, **106–107**
Po River, location, **82–83**
Pohai, Gulf, 10–11
Poljes, Dalmatia, 105
Pomeranian lake region, Germany, 136
Ponchartrain, Lake, **139**
Ponds, Nebraska, **133**
Portland, Bill of, England, **34–35**
Potash, in Nebraska lakes, 133
Potomac River, 79, 110–111; Great Falls, **118–119**
Presidential Range, 95
Prevailing westerlies, 45
Puerto Rico, **60–61**
Purbeck Downs, England, **29**
Pyrenees, 13, 96–97

Quaternary deposits, 83

"Racetrack," Black Hills, **101**
Radial rivers, **94–95, 96–97**
Rainier, Mount, **94–95**
Rapidan River, 110–111
Rappahannock River, 110–111
Raquette Lake, Adirondacks, **93**
Raritan Bay, **77**
Rectangular islands, **60–61**; lakes, **124–125**; rivers, **92–93**
Red Valley, Black Hills, **101**
Rejuvenation of rivers, 109
Rhine-Marne Canal, 98
Rías, Spanish coast, 43
Ribbon Falls, Yosemite, 120–121
Ridges, folded mountain, **109**
Rift valleys, 147
Ring-like, or annular, rivers, **100–101**
River embayments, Delaware and Chesapeake bays, **77**
Rivers: annular, **100-101**; antecedent, 113; centripetal, **98–99**; dendritic, **102–103**; general discussion, 76; interrupted, **104–105, 106–107**; parallel, **84–85, 86–87, 88–89**; piedmont, **82–83**; radial, **96–97**; rectangular, **92–93**; superposed, 111; trellis, **90–91**
Roanoke River, 110–111
Rock falls, Palisades, 27
Rockaway Beach, Long Island, **32–33, 40–41**
Rocks, metamorphic, 51
Roman roads, 48
Ronkonkoma Lake, Long Island, 55; moraine, 55
Rotorua geyser region, New Zealand, 114
Ruapahu volcano, New Zealand, 15
Rügen Island, Baltic Sea, 6
Ryukyu Islands, Pacific, 64

Saba, West Indies, 67
Sable Island, Nova Scotia, **74–75**
Sacré Coeur, Paris, 151
St. Kitts, West Indies, 66–67
St. Lucia, West Indies, 66–67
St. Vincent, West Indies, 66–67
Salmon River Mountains, Idaho, 103
Salt Lake City, 97
Salt lakes and flats, 107
Salton Sea, California, **138–139**
Samoa, 64
San Joaquin River, **82–83**
Sand and gravel, in alluvial fans, 107; on beaches, 31, 41
Sand Hill Region, Nebraska, 133
Sandy Hook, **33**
Sardinia, 34–35, **60–61**
Saugeen (Indian) Peninsula, Ontario, **18–19**
Scandinavia, glacial lakes, 126